RECENT ADVANCES IN PLASTICIZERS

Edited by **Mohammad Luqman**

Recent Advances in Plasticizers
Edited by Mohammad Luqman

Published by InTech
Janeza Trdine 9, 51000 Rijeka, Croatia

Publishing Process Manager Iva Simcic

Technical Editor Teodora Smiljanic

Cover Designer InTech Design Team

Additional hard copies can be obtained from orders@intechopen.com

Recent Advances in Plasticizers, Edited by Mohammad Luqman
p. cm.
ISBN 978-953-51-0363-9

For EU product safety concerns:
IN TECH d.o.o., Prolaz Marije Krucifikse Kozulić 3, 51000 Rijeka, Croatia,
info@intechopen.com or visit our website at intechopen.com.

Meet the editor

Dr Mohammad Luqman is serving as an Assistant Professor in the Chemical Engineering Department, King Saud University (KSU), Saudi Arabia. He teaches Polymer Science and Engineering courses. He served as an Assistant Manager in Samsung Cheil Industries, South Korea, and worked on the development of heat resistant polymers, organic glass, and block copolymers as impact modifiers and compatibilizers for engineering polymers. He served as a post-doctoral fellow at Artificial Muscle Research Center, Konkuk University, Korea. He earned his PhD in the field of Ionomers from Chosun University, Korea. Dr Luqman studied extensively the effects of various types of additives/plasticizers on the morphology and dynamic mechanical properties of ionomers. Dr Luqman is acting as a Regional (Middle East) Editor to an International Journal "Chemical and Environmental Research" and a Section Editor to "Journal of Industrial Research & Technology".

Contents

Preface

Plasticizers are a class of materials which are added not only to plastics but to cement, clay, concrete, wallboard, etc., to increase the process-ability, flexibility and durability, and of course to reduce the cost in many cases. This book deals with the plasticizers used in plastic, electrochemical, environmental, and health sectors.

This edition is comprised of 10 chapters. Without plasticizers, it is very hard to get the desirable properties in polymers. However, they, mainly phthalate plasticizers, pose serious threat to environment and health. Taking these concerns into account, the last few decades have seen tremendous efforts in finding non-toxic, low cost, and effective alternatives of the existing plasticizers. In this regard, the first chapter introduces the scientific community to an innovative and breakthrough technology in the field of PVC plasticizers, by discussing in detail about Flexidone™ plasticizers; free from physiological concerns. Saving our nature and caring for the environment is highly desirable. Hopefully, chapter 2 will be very interesting to the readers as it provides a detailed information about a new source of Collagen based bio-plasticizers that can be obtained from discarded materials from a natural source; Jumbo Squid (*Dosidicus gigas*).

The third chapter is about pharmaceutically used plasticizers. Authors have discussed in significant detail about the types, roles, nature, selection criterion, etc., of plasticizers for pharmaceutical applications. In chapter four, the authors have presented the use of plasticized polymers in pharmaceutical/bio-fields, including, as coatings and films of pharmaceutical dosage forms, membranes for transdermal drug delivery systems, in-situ implants, matrix polymer systems, and bio-adhesives. Transdermal drug delivery method is one of the non-destructive methods of drug administration. In chapter 5, authors have enlightened us about various aspects of plasticizers and transdermal drug delivery systems including the role and effectiveness of plasticizers in these systems, compositions and types of these systems, and the current developments related to transdermal formulations.

In the last few decades, the ion-selective electrodes/electrochemical sensors have been widely studied and applied in various analytical projects. The electrodes based on plastics are one of various types of electrodes, e.g. solid state electrodes, liquid membrane electrodes, and gas membrane electrodes. Plasticizers are one of the major components of these plastic membranes. Chapters 6 and 7 discuss roles of plasticizers for the same.

Conventional liquid electrolytes have few drawbacks including leakage of hazardous liquids or gases, formation of lithium dendrite, electrolytic degradation of electrolyte, and exhibiting short-term stability due to the evaporation of the liquid phase in the cells. To address this issue, in chapter 8, authors have discussed the preparation and characterization of high molecular weight poly(vinyl chloride)–lithium tetraborate electrolyte, a solid polymer electrolytes (SPEs), plasticized by propylene carbonate.

The effect of concentration, and type of plasticizers (phthalate and non-phthalate based), on the mechanical properties of cellulose acetate butyrate organic-inorganic hybrids synthesized via sol-gel is presented in chapter 9. The author has presented tributhyl citrate (TBC) and triethyl 2-acetylcitrate (TEA) as an alternative to phthalate based plasticizers. The last chapter is very important which deals with drawbacks of plasticizers. Authors discussed the health risk assessment of phthalate esters in wastewater effluents and receiving freshwater systems.

This book is the result of hard work and dedication of the authors who tried their best to present the latest information and shared their experience in the field. The timely support from publishing process managers are thankfully acknowledged. The very friendly behavior, support, encouragement and the lively environment provided by the departmental colleagues, more specifically the Chairman of the Chemical Engineering Department, King Saud University, is gratefully acknowledged.

Mohammad Luqman
Assistant Professor
Chemical Engineering Department,
College of Engineering, King Saud University,
Kingdom of Saudi Arabia

1

Flexidone™ – A New Class of Innovative PVC Plasticizers

Martin Bonnet[1] and Hasan Kaytan[2]
[1]*University of Applied Sciences Cologne*
[2]*ISP Global Technologies Deutschland GmbH*
Ashland Specialty Ingredients
Germany

1. Introduction

It has been found, that N-alkyl-(C8 to C18) pyrrolidones are highly efficient, strong solvating performance plasticizers which decrease gelling temperatures substantially (Bonnet & Kaytan, 2008). They facilitate fast gelling, produce flexibility at extremely low temperatures. Higher alkyl pyrrolidones also exhibit very low volatility. After these excellent properties were proven by industrial trials in several different applications such as flooring, gloves, window sealing and wires, they were introduced to the market as Flexidone plasticizers.

Solubility temperatures of the N-alkyl-pyrrolidones (DIN 53408) are between 52°C (N-octyl-pyrrolidone) and 80°C (N-octadecyl-pyrrolidone). Accordingly, the gelling temperatures are substantially lower than with the standard plasticizers.

Plasticizing efficiency of the different Flexidone Types were tested through comparative determination of Shore A values. It could be shown that Flexidones are about 30-50% more efficient than the standard plasticizer DINP (diisononyl phthalate).

Cold Flexibility with the Flexidones was checked by the Folding test DIN EN 495-5.

Further trials in filled systems showed that Flexidones are highly compatible with e.g. calcium carbonate and allow very high filler loads with outstanding mechanical properties. Manufacturing tests with a highly filled system using an extruder resulted in increased output while significantly reducing plasticizer levels and processing temperatures.

In the broader effort to offer to the flexible-PVC industry cost-effective plasticizer systems with desired process- and end product properties, tests were performed on blends with various low cost secondary plasticisers. In these experiments, all Flexidone types have worked as performance boosters and compatibilizers.

Exemplarily the results of blends with certain fatty acid esters and chlorinated types will be presented. As a function of the ratio of mixture indentation hardness, tensile properties and gelling properties (plastisol) have been evaluated.

Compared to common systems, these new Flexidone mixtures surpass the performance characteristics and are superior to most of the phthalate and phthalate free systems. All of the products are now industrially produced REACH-compliant types that are globally available in appropriate volumes.

2. Properties of Flexidones in soft-PVC applications

In terms of worldwide consumption polyvinyl chloride (PVC) stands in third place behind polyethylene (PE) and polypropylene (PP). Thank to the development of a wide range of functional additives, in particular thank to effective plasticizers PVC could achieve this important commercial relevance. PVC is one of the few thermoplastics whose hardness can be adjusted from rubber-like elasticity up to hard formulations (Franck & Knoblauch, 2005). Thus for well over 50 years plasticizers have been playing a significant role in the manufacture of soft PVC products for the most versatile applications from floor coverings to roof membranes, cable insulation to blood bags.

With a market share of approximately 85 % phthalate plasticizers – di-2-ethyl hexyl phthalate (DEHP), diisononyl phthalate (DINP) and diisodecyl phthalate (DIDP) – represent the most significant class of plasticizers at present. They are the all-rounders amongst plasticizers. The remaining 15 % are taken up by plasticizers that show excellent properties in particular areas even if they have weaknesses in others. For instance trimellittic acid esters exhibit particularly good heat stability, whilst phosphoric acid esters confer fire resistance. Polymeric plasticizers (polyesters) come into play when excellent oil resistance and very good migration behaviour is required.

However, for more than 25 years, plasticizers in particular phthalates, have been the subject of environmental and health debate despite attempts by the industry to defend their current status with ever new data. This has however initiated the development of numerous new plasticizers as phthalate substitutes with less toxicological concern. One of the best known examples of these new plasticizers is Hexamoll DINCH (1.2-cyclohexanedicarboxylic acid diisononyl ester) which was developed by BASF for sensitive applications.

Unfortunately most of the newly developed plasticizer alternatives do hardly offer any improvements in the processing behaviour or property profile of soft PVC alongside the ecological or toxicological factors.

Now, however, ISP-Ashland Specialty Ingredients, Cologne, Germany, and the Institute for Materials Technology at the University of Applied Sciences Cologne (Institut für Werkstoffanwendung der Fachhochschule Köln) have collaborated to develop a new class of plasticizers for PVC based on linear alkyl pyrrolidones. Initial results show that they are not only free from physiological concerns - e.g. acute toxicity is relatively low, dependent on alkyl chain length so that the LD_{50} for example lies between 2.05 g/kg for Flexidone 100 (C-8 Pyrrolidone) and >12 g/kg for Flexidone 500 (C-16/18 Pyrrolidone) (Ansell & Fowler, 1988) –, but also possess several outstanding properties. These properties enable gentler, cost saving processing of soft PVC and make it possible to produce highly flexible products for low temperature applications.

2.1 Structure and mode of action

Due to the planar structure of pyrrolidones the oxygen with its high electronegativity can easily cause an electron to delocalize (Fig. 1). This produces a strong dipole moment.

Chemically binding a flexible non-polar alkyl chain with a compact hydrophilic head makes the alkyl pyrrolidones soluble in both polar and non-polar solvents. Even though the alkyl chain length can be adjusted to lie between C4 and C30 it has been found that chain lengths between C8 and C18 are particularly suitable for use as plasticizers. Due to their excellent dissolving power and good compatibility with PVC both gelling temperature (very low solubility temperatures - see Fig. 2) and gelling time (see gelling curves Fig. 9 and 14) can be substantially reduced. At the same time this leads to highly flexible PVC formulations that do not lose their flexibility even at extremely low temperatures.

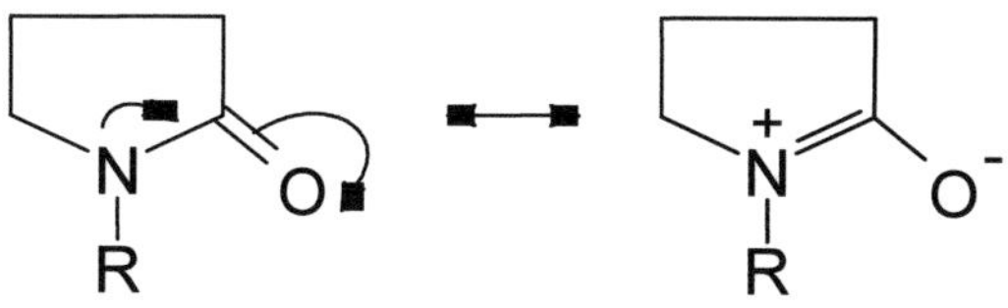

Fig. 1. Electron delocalization in pyrrolidones.

2.2 More cost effective dry blending

A measure of the effectiveness of a plasticizer is its solubility temperature. This is the temperature at which a plasticizer completely dissolves a given PVC. Typical solubility temperatures lie between 87°C for butyl benzyl phthalate and 151°C for DINCH, with diisononyl phthalate at 129°C. Figure 2 shows the solubility temperatures for alkyl pyrrolidones with various alkyl chain lengths. It can be seen that the solubility temperature can be adjusted by chain length to lie between 52 and 80°C and is thus significantly lower than the solubility temperatures of conventional plasticizers.

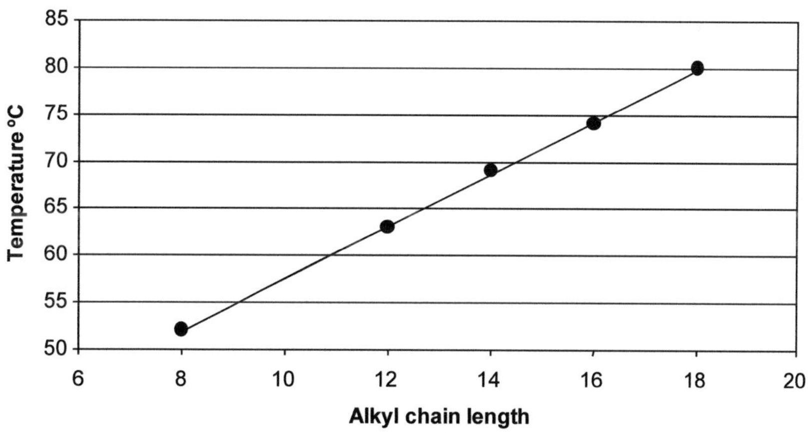

Fig. 2. The dependence of solubility temperature of various alkyl pyrrolidones on alkyl chain length in accordance with DIN 53 408.

This also reduces the time taken to produce a dry blend in a high speed cooler mixer without the need for external heating. Depending on the formulation mixing times can be reduced down to 20 % in comparison to phthalate plasticizer formulations. In addition, processing temperatures are lowered by 20 to 40°C in comparison to classic soft PVC processing. These significantly lower temperatures allow the use of temperature sensitive additives such as special colorants and scents and result in clear time and cost savings through the use of alkyl pyrrolidone plasticizers in comparison to standard plasticizers.

2.3 Cold break at temperatures lower than -70°C

The efficiency of the plasticizing effect of the Flexidones can be seen very clearly in comparative measurements of hardness (Shore A) in relation to the plasticizer content with DINP as the standard plasticizer (Fig. 3). These show e.g. a hardness of 80 Shore A can be achieved with 33 parts of Flexidone 300 (C-12 Pyrrolidone) compared to 60 parts of DINP. In this example the same flexibility can be reached with 45 % less plasticizer.

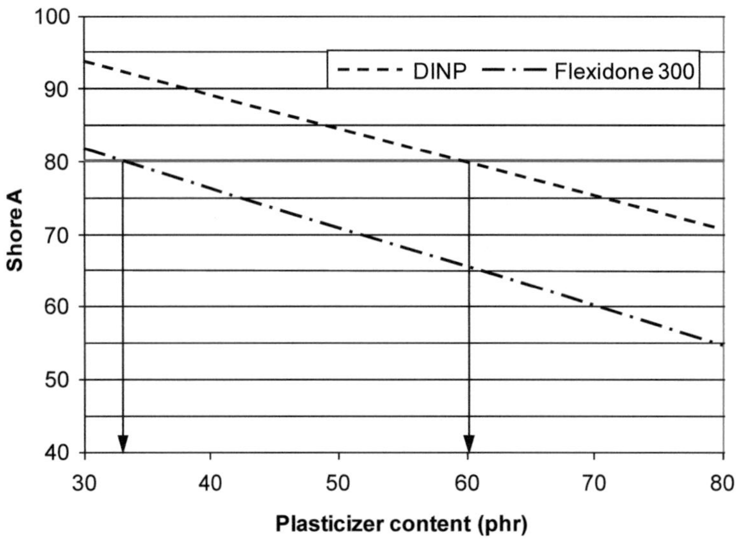

Fig. 3. The dependence of hardness on plasticizer content (parts per100 parts of PVC) tor DINP and Flexidone 300.

The hardness is determined at room temperature, however, in many applications the temperature can be temporarily or permanently significantly lower. Many soft PVC formulations are not only much harder at lower temperatures but also completely lose their toughness so that they are subject to brittle fracture under flexural or tensile loadings. A practical test for determining this boundary temperature is DIN EN 495-5 (foldability at low temperature). However, since it is not only the plasticizer type and concentration that is responsible for low temperature behaviour, but also the molecular weight of the PVC grade used, tests were performed with two different concentrations (40 and 60 parts) of Flexidone 300, Flexidone 500 and DINP in PVC grades with K-values of 60, 70, 80 and 99 (Fig. 4). This showed that the cold break temperature could be reduced by 15 to 30°C through the use of Flexidones. At 60 parts of Flexidone 300 and 500 the exact cold break temperature could not be determined for the higher K-values since the cooling system of the tests apparatus could only produce temperatures down to -70°C and at this temperature none of the samples with each 60phr Flexidone 300 and 500 showed breakages or cracks. Flexidone 500 is therefore the first available low volatility plasticizer which could facilitate cold break temperatures of lower than -70°C.

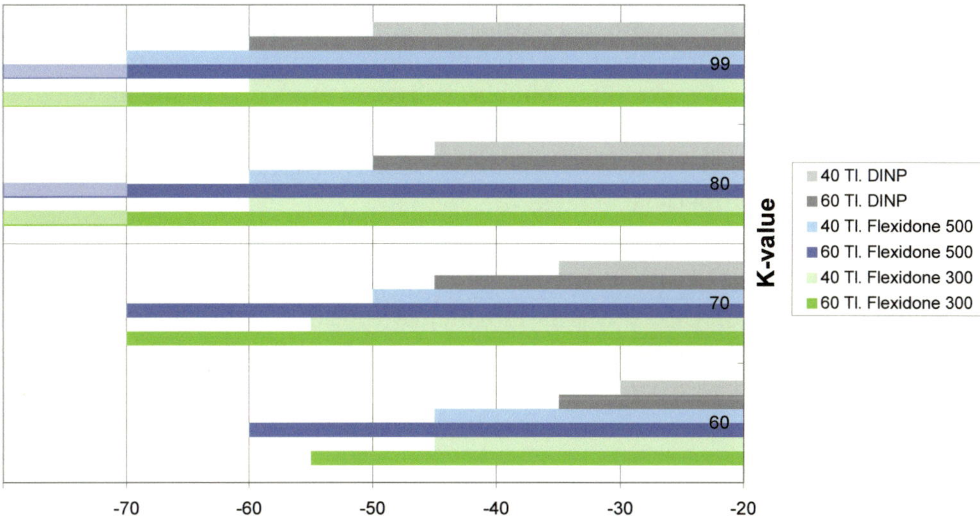

Fig. 4. Determination of the low temperature break behaviour according to DIN EN 495-5 (foldability at low temperatures) for soft PVC sheets with various concentrations of Flexidone 300, Flexidone 500 and DINP for a range of PVC grades with different K-values.

Figure 5 shows the progression in Shore A hardness at temperatures between +20°C and -50°C for Flexidone 300, Flexidone 500 and DINP as well as DOA which is a widely used low temperature plasticizer. It is very noticeable that Flexidone grades not only have a better

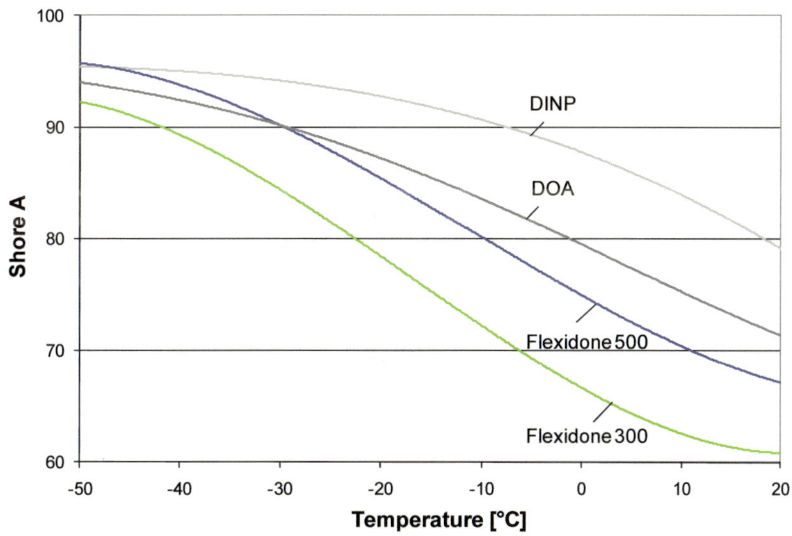

Fig. 5. Variation of Shore A values at temperatures between +20°C and -50°C for samples with 60 parts of Flexidone 300, 500, DOA (dioctyl adipate) and DINP.

plasticizing power in comparison to DINP and DOA at room temperature, but the progression in hardness with decreasing temperature is also very different, i.e. an initially even gradient is followed by a sharper rise. This means that Flexidones show significantly higher cold flexibility than conventional plasticizers at temperatures in the region of -20 °C which are typical for exterior applications. Thus the use of Flexidone grades not only enable extremely low cold break temperatures, but also delivers soft PVC with significantly better flexibility at very low temperatures.

2.4 Good mechanical properties

As Figures 6 and 7 show the mechanical properties of soft PVC film made with Flexidone 500, particularly at higher K-values, have comparable mechanical values to samples made with the same content of DINP. This is remarkable considering that these samples are significantly softer whilst as shown earlier also retaining their flexibility down to the very low temperatures.

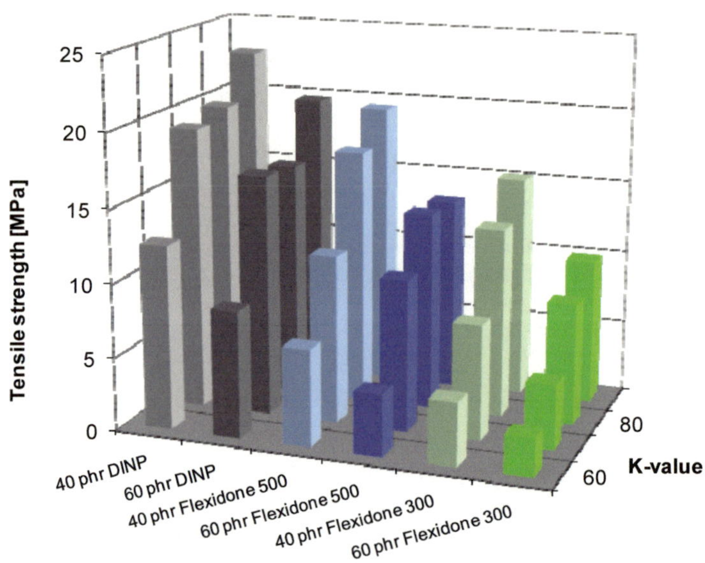

Fig. 6. Tensile strength of flexible PVC sheets with various concentrations of Flexidone 300, Flexidone 500 and DINP in PVC grades with different K-values.

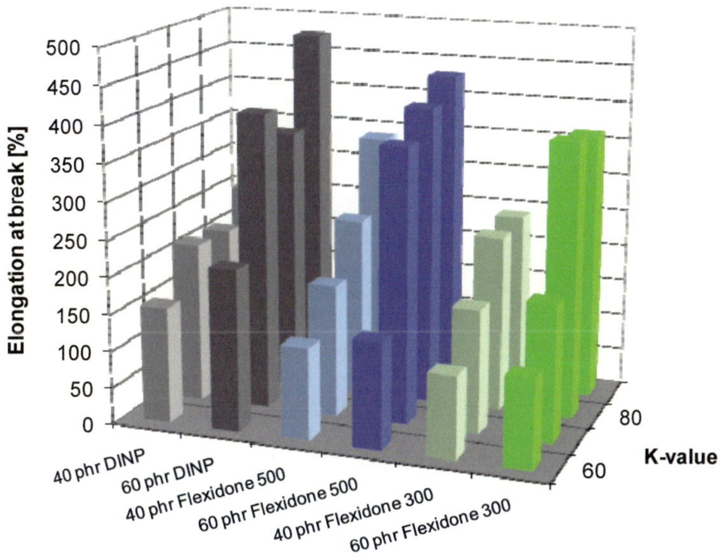

Fig. 7. Elongation at break for soft PVC sheets with various concentrations of Flexidone 300, Flexidone 500 and DINP in PVC of different K-values.

2.5 Properties of highly filled systems

Fillers are used in plasticized PVC to reduce costs, and also to facilitate a special change in properties so that the compound largely meets the requirements of the end products (Hohenberger, 2001). Among the fillers, calcium carbonate, with a worldwide market share of approximately 70%, plays the dominant role. For plasticized PVC, depending on the application (cable, floor covering, profiles, films), uncoated or coated calcium carbonate grades, with different particle sizes, can be used. For all experiments a stearic acid coated calcium carbonate with a d50% value of 2.4 µm and a top cut of 20 µm (Omya BSH) was used.

In highly filled systems, e.g. with a calcium carbonate content of 150 phr important mechanical properties, like the tensile strength, can easily drop down to less than 50%, compared to the unfilled systems if a standard plasticizer like DINP is used. For a highly filled system with Flexidone 300 as plasticizer, all essential mechanical properties vary only insignificantly from the unfilled system. As shown in Figure 8, softness of the Flexidone 300 formulation decreases only slightly in comparison to DINP even with a filler load of 100 phr Calcium carbonate.

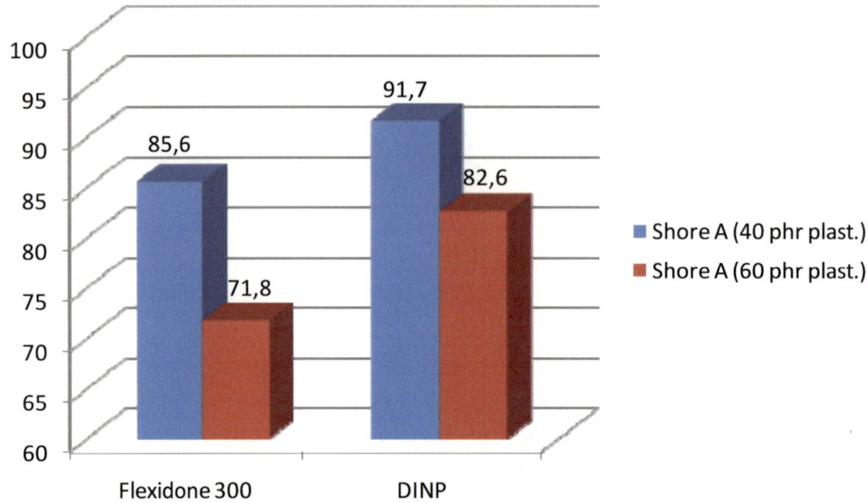

Fig. 8. Shore A-values with Flexidone 300 and DINP, each filled with 100 phr $CaCO_3$.

Therefore the formulation costs can be lowered in many applications using Flexidones with remarkably increased filler loads.

2.6 Improvements in plastisol production

Plastisols take a special place in processing of PVC since they are moulded as liquids or pastes rather than in a thermoplastic state. The solidification, the so called gelling, takes place at the end of molding through heat treatment at 120 to 200°C (Franck & Knoblauch, 2005).

In making these pastes the processing behaviour can be significantly improved in many respects by the substitution of around 10% of the plasticizer with Flexidone 100 (octyl pyrrolidone). Degassing at the end of mixing and homogenizing the plastisol mixture is intended to minimize defects in the subsequent processing of the pastes. Flexidone 100 is amongst the surfactants with the highest dynamic wetting properties. This means that a partial substitution is sufficient in order to reduce foaming to a minimum and thereby significantly shorten degassing cycles.

During the processing of PVC plastisols the gelling temperature is of particular interest. Therefore substantial efforts have been made in order to lower the gelling temperature. It is determined via measurements of the complex viscosity against temperature. After an initial drop once gelling begins there is a rise in the viscosity of more than four orders of magnitude. The gelling temperature is the point of inflection on the curve. Figure 9 shows viscosity measurements of plastisols with 50 parts Vestinol 9 (DINP) and plastisols, in which 1, 3, 5 and 10 parts of Vestinol 9 were replaced by Flexidone 100. Due to the significantly higher dissolving power of Flexidone 100 the gelling temperature can be significantly reduced. Each part of Flexidone 100 causes approximately 2°C drop so that a blend of Vestinol 9/Flexidone 100 at a ratio of 40/10 reduces the gel temperature by 20°C. The viscosity profile is at the same time not affected at all and only after storage for a long time a small increase in viscosity could be seen.

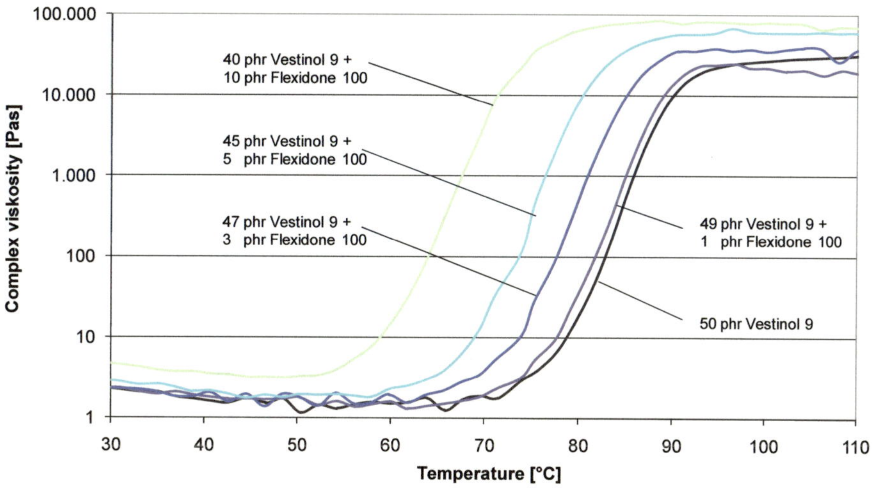

Fig. 9. Gelling curves of PVC plastisols (50 phr plasticizer) with various mixtures of Vestinol 9 (registered trade mark of Evonik for DINP)/Flexidone100.

Comparable results were also found for blends with Hexamoll DINCH, but the initial viscosities were much lower and the viscosity after prolonged storage rose insignificantly.

3. Mixed plasticizer systems

Through these outstanding processing and physical properties and the extremely high compatibility, Flexidones enhance the tolerance level of low cost secondary plasticizers in PVC so that these could be used in very high amounts. Adding Flexidone 300 or 500 to secondary plasticizers like fatty acid esters, chlorinated Paraffines/esters, ESO (epoxidized soybean oil) as well as to primary plasticizers like DOA

decreases the

- indentation hardness
- cold flexibility temperature
- processing temperature
- processing time

and increases the

- compatibility
- gelling speed
- clarity / transparency.

3.1 Flexidone-fatty acid ester-system

In figure 10 and 11 the effect of Flexidone 300 and 500 on the tensile strength and elongation at break in mixtures with a fatty acid ester can be seen. All mixtures have a total plasticizer content of 60 phr. The results prove the excellent compatibility of this system with good mechanical properties in all concentrations up to 75% fatty acid ester.

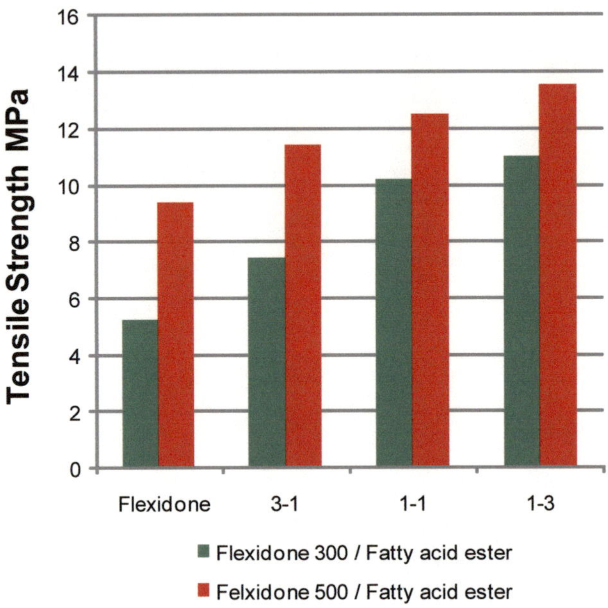

Fig. 10. Tensile strength for Flexidone 300 and Flexidone 500 in mixtures with a fatty acid ester.

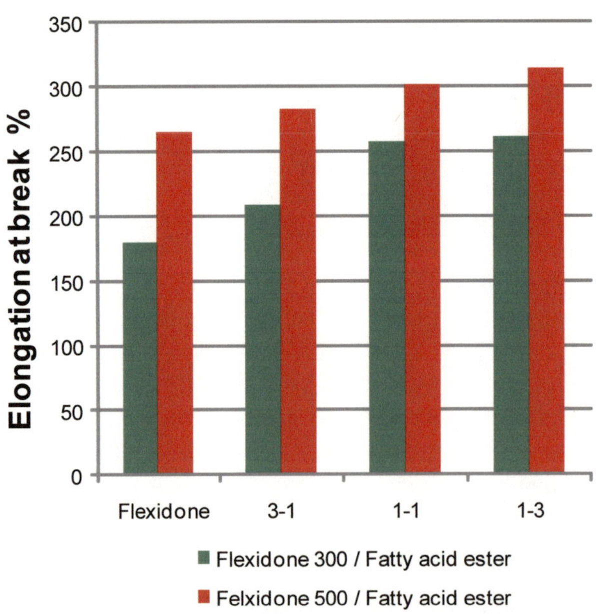

Fig. 11. Elongation at break for Flexidone 300 and Flexidone 500 in mixtures with a fatty acid ester.

As it can be seen in figure 12 the hardness increases with increasing fatty acid ester content. But all examined mixtures show still higher plasticizing efficiency than DINP.

Most interestingly all mixtures with a ratio between 3:1 and 1:3 show cold foldability temperatures of -70°C and below! Therefore the costs can be reduced by mixing Flexidone's with fatty acid esters while improving tensile properties and still benefit from excellent cold flexibility.

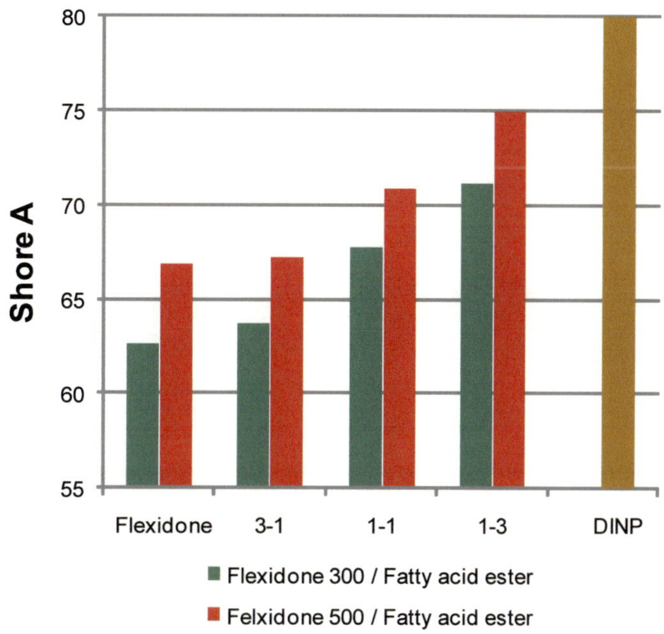

Fig. 12. Indentation hardness for Flexidone 300 and Flexidone 500 in mixtures with a fatty acid ester in comparison to DINP.

As a conclusion of these results ISP commercialized different Flexidone / fatty acid ester - blends such as Flexidone 350FE, Flexidone 333FE, Flexidone 550FE and Flexidone 533FE. Flexidone FE yields stable, low-viscosity plastisols with excellent shelf life and desirable fast-fusing properties. This combination enables the system to be used in various production methods such as extrusion, calendaring and injection molding. Additionally they provide low-viscosity plastisols with good gelling properties. As figure 13 indicates, of various samples of plastisol formulations, those with Flexidone 333FE and 533FE exhibit the best viscosity stability after 28 days, enlarging the processing window for many product manufacturers.

In addition to better viscosity stability, Flexidone FE improves the gelling behaviour of PVC products, which affects its strength. In figure 14, the gelling rates of Flexidone 333FE and 350FE were the fastest compared with other Flexidone grades and a DINP.

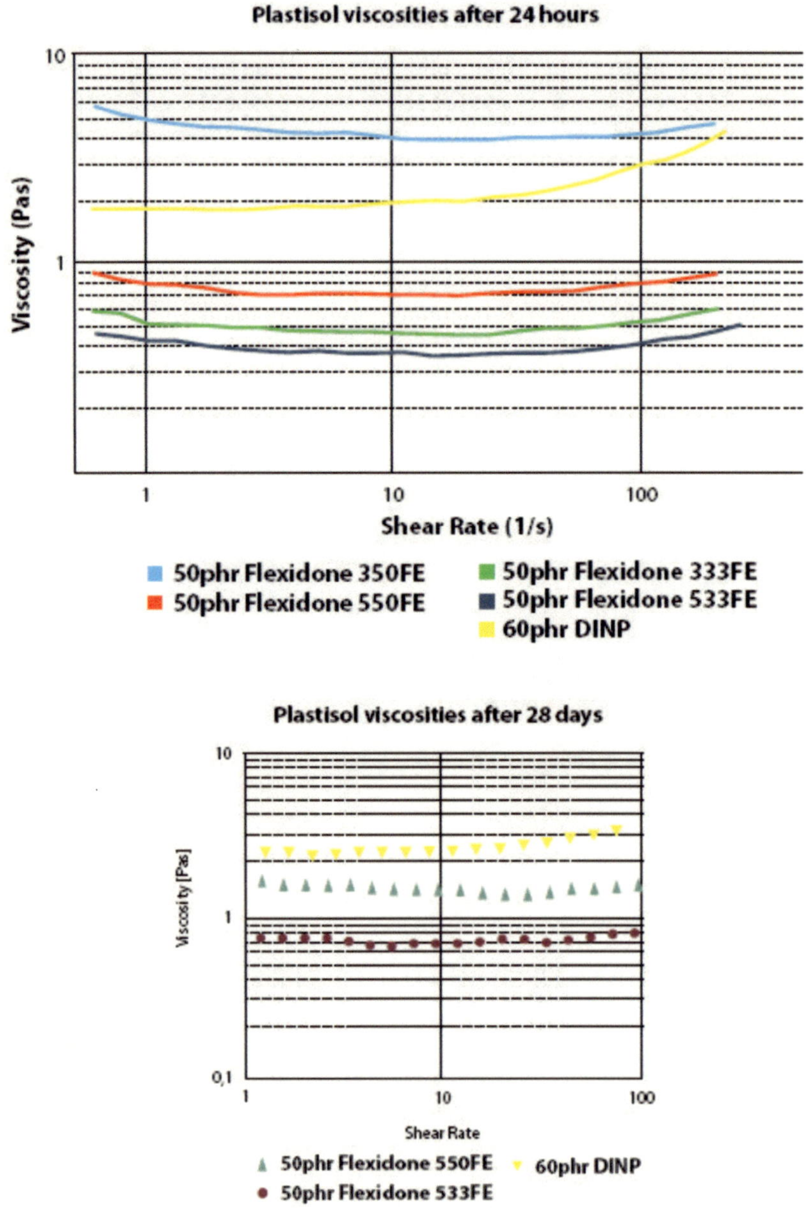

Fig. 13. Viscosity of various plastisol formulations with Flexidone FE Grades compared with DINP.

The Flexidone FE grades behaviour in plastisol can be summed up as:

- Flexidone 350FE offers fast and low-temperature gelling, improved transparency and homogeneity. As a primary plasticizer for plastisols (for immediate processing), it is an extremely fast fusing system requiring about 30% less use than standard plasticizers.

- Flexidone 333FE can be used as primary plasticizer; about 15% more efficient than other systems; has lower plastisol viscosity than 350FE and stays stable even after longer storage.
- Flexidone 550FE can be used as primary plasticizer with gelling properties similar to 333FE but at a lower viscosity with higher clarity and lower volatility.
- Flexidone 533FE for very-low-viscosity plastisols; can also be stored for very long time. Lower volatile.

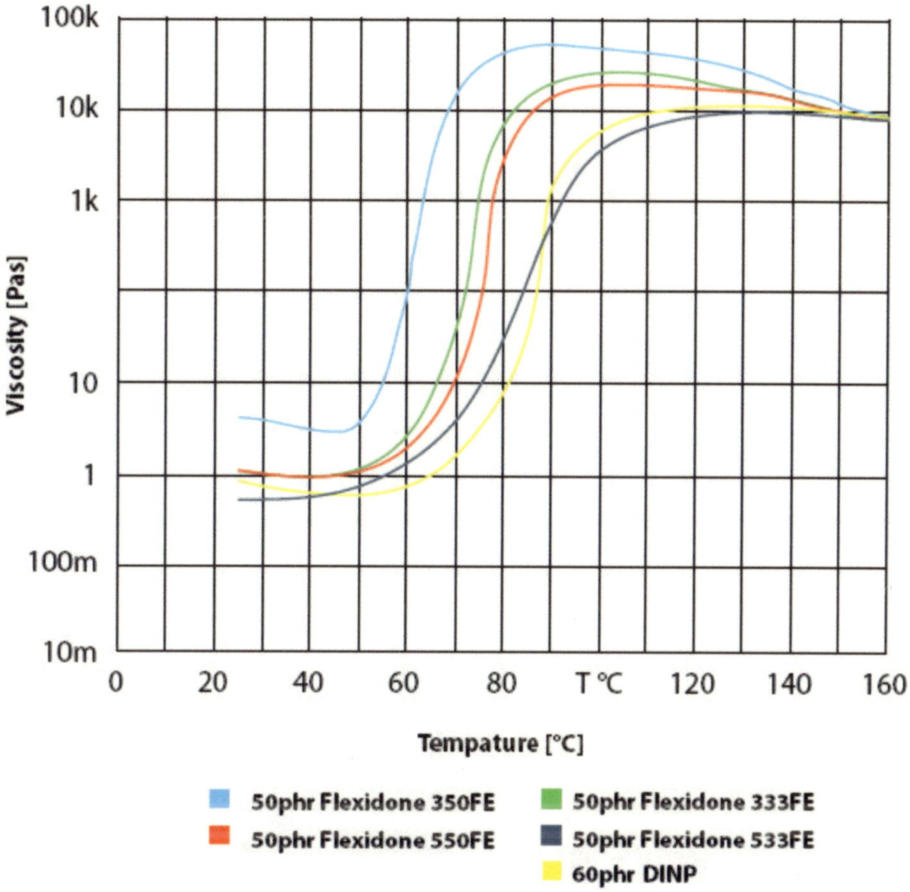

Fig. 14. Gelling curves of Flexidone FE Grades and DINP.

3.2 Flexidone-chlorinated ester-system

Similar experiments as with the Flexidone/fatty acid ester-systems were performed for the plasticizer systems Flexidone/chlorinated ester. In figure 15 and 16 the effect of Flexidone 300 and 500 on the tensile strength and elongation at break in mixtures with a Cl-ester can be seen. All mixtures have a total plasticizer content of 60 phr.

The results prove the superior compatibility of this system with good mechanical properties in all concentrations up to 75% Cl-ester. Especially the 1:1-mixture with Flexidone results in a rubber like behaviour with elongation at break values of over 500%. From this 500% elongation over 450% is elastic deformation and less than 50% is plastic deformation.

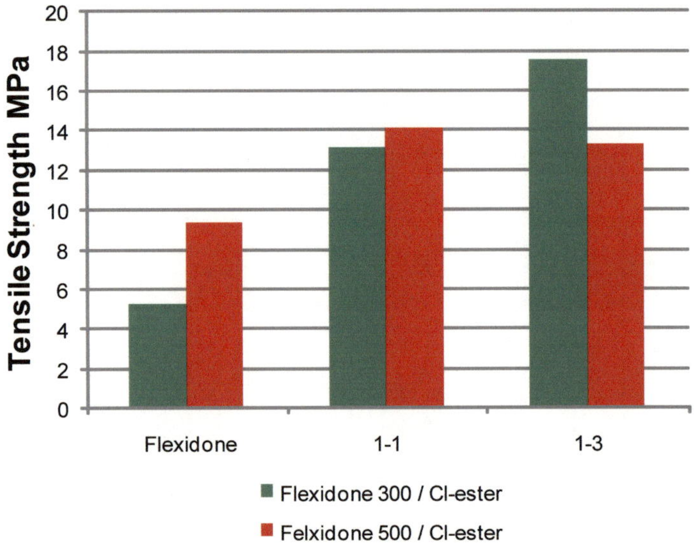

Fig. 15. Tensile strength for Flexidone 300 and Flexidone 500 in mixtures with a Cl-ester.

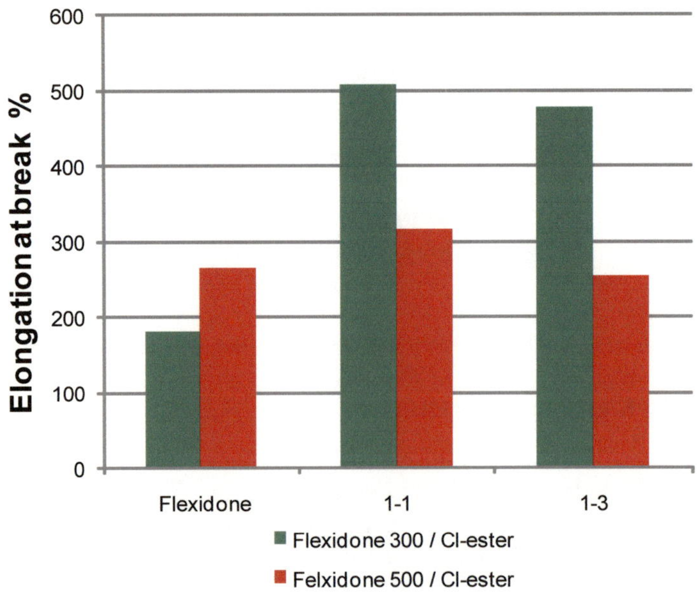

Fig. 16. Elongation at break for Flexidone 300 and Flexidone 500 in mixtures with a Cl-ester.

Although the plasticizing efficiency reduces with increasing Cl-ester content, even Flexidone/Cl-ester-mixtures with very high Cl-ester content show Shore A values similar to a sample with the same amount of DINP (see figure 17).

Analogue to the indentation hardness results the cold flexibility in the mixtures with Cl-ester is slightly inferior to the corresponding Flexidone/fatty acid ester system. Still all measured samples show cold foldability temperatures of -40°C and below.

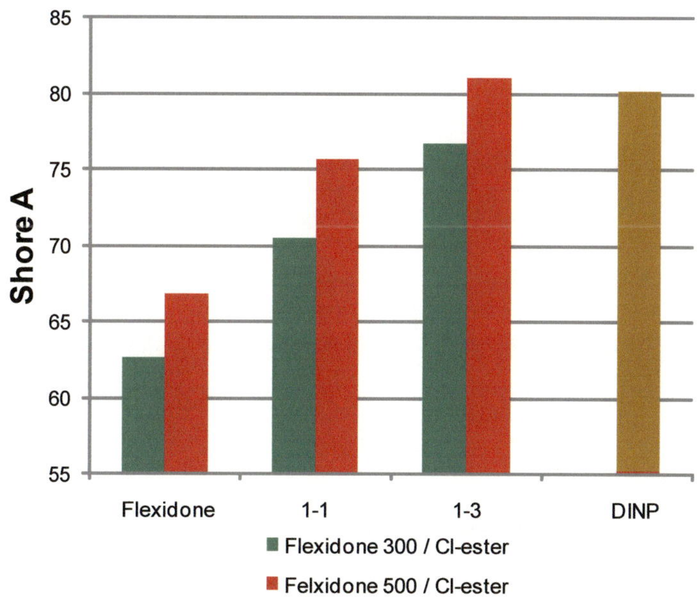

Fig. 17. Indentation hardness for Flexidone 300 and Flexidone 500 in mixtures with a Cl-ester in comparison to DINP.

4. Product developments

Over the last few years a significant number of product trials have been performed to test Flexidone in existing formulations or in new product developments successfully.

4.1 Shoe soles

Shoe soles are just one example for various soft-PVC applications ISP was asked to come up with alternative formulations. Following two formulations will be presented here:

- one formulation with 80 phr DINP, 40 phr chlorine paraffin and 20 phr $CaCO_3$ and
- one formulation with 80 phr DOA.

As a substitute for the DINP/CP-system (with Shore A 53) we could offer a Flexidone/ESO system with 80 phr ESO and 20 phr Flexidone 300 showing the same softness (Shore A 50) and lower processing temperature even with a lower plasticizer content and an increased amount of $CaCO_3$ (see Table 1).

PVC	100	100
DINP	80	-
CP	40	-
Flexidone 300	-	20
ESO	5	80
Plasticizer total	125	100
$CaCO_3$	20	50
Stabilizer	2,5	2,5
Shore A	*53*	*50*

Table 1. Formulation and Shore A for soft-PVC shoe soles containing Cl-paraffin.

In case of the DOA system the aim was to improve the softness from Shore A 60 to 56, at the same time increasing of the compatibility/proccessability of DOA and improving the clarity. The target could be achieved by replacing 20 phr DOA with the same amount of Flexidone 300 (see Table 2).

PVC	100	100
DOA	80	60
Flexidone 300	-	20
ESO	5	5
Stabilizer	2,5	2,5
Shore A	*60*	*56*

Table 2. Formulations and Shore A for soft-PVC shoe soles with reduced Hardness.

These two examples show what a powerful tool Flexidone is in solving a lot of practical processing and/or performance problems of flexible PVC formulators. An extensive set of experimental data is available to help finding solutions to meet customers' needs.

4.2 Avoiding use of critical plasticizers

In studies of rodents exposed to certain phthalates, high doses have been shown to change hormone levels and cause birth defects (National report, 2009). Therefore in the U.S. children's toy or child care article that contains concentrations of more than 0.1 percent of DEHP, DBP (dibutyl phthalate), or BBP (butylbenzyl phthalate) are illegal (Congress, 2007). In plastisol applications DBP and BBP are widely used as fast-fusing plasticizers in DINP. As figure 18 proves, Flexidone 100 and 300 can easily replace fast-fusers with unfavourable ESH profile.

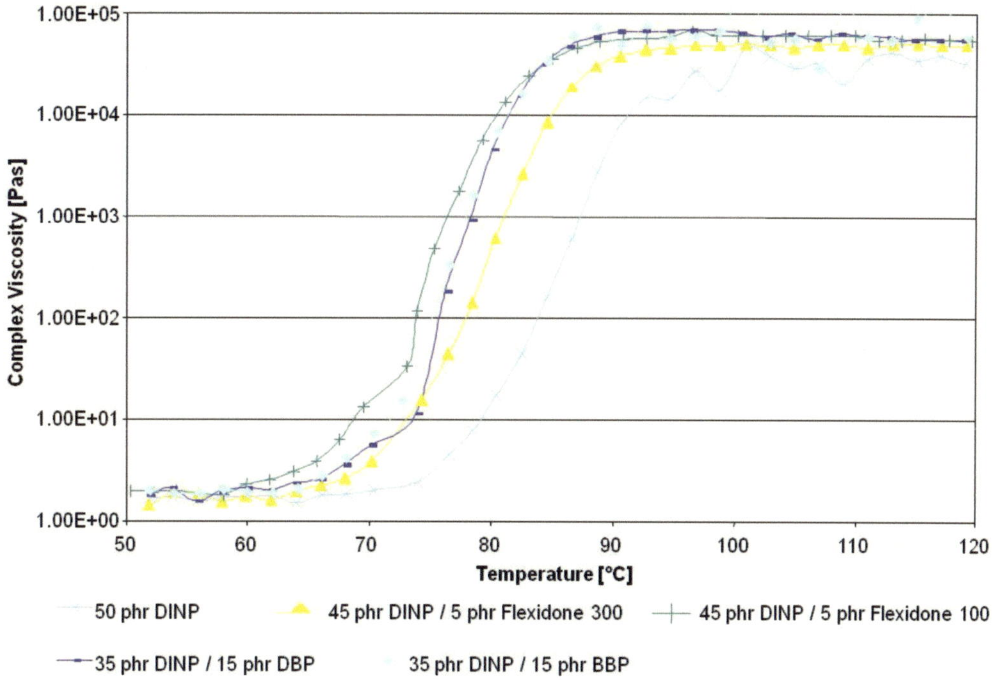

Fig. 18. Gelling curves of PVC plastisols (50 phr plasticizer) with various mixtures of Vestinol 9 and Flexidone100, 300, DBP and BBP.

5. Conclusion

It could be shown, that Flexidone's are very efficient PVC plasticizers. They are up to 50% more efficient than standard plasticizer. Since they improve a lot of mechanical and processing properties in mixed plasticizer systems, cost effective plasticizer systems with superior properties could be presented. In any case the presence of Flexidone in the plasticizer system results in

- lower plasticizer use (⇨ easier processing and less migration)
- lower gelling temperatures (⇨ energy saving)
- faster gelling (⇨ higher production rates)
- better cold flexibility (⇨ more durable products at low temperatures)
- better compatibility (⇨ higher transparency and homogeneity as well as more fillers and use of less compatible materials)

Therefore the Flexidone family of plasticizers represents breakthrough technology in "cold flex" performance of plastics. Depending on the application, it can offer a great deal more. Flexidone plasticizers expand the options in PVC formulation, manufacturing and end-product design. This flexibility lets re-imagine the potential of demanding applications and reconfigure processing for unprecedented efficiencies.

6. References

Kaytan, H., Bonnet, M. (2008), *New, innovative PVC plasticizers: N-alkyl-pyrrolidones*, in The 10th International PVC Conference, IOM Communications Ltd (2008), p. 305

Kaytan, H., Bonnet, M. (2008), *N-alkyl pyrrolidones as innovative PVC plasticisers*, Plastics, Rubber and Composites, 37 (2008) 9/10, p. 411

Bonnet, M., Kaytan (2008), H., *Flexible Even at Very Low Temperatures*, Kunststoffe international, 12 (2008), p. 62-65

Frank, A. & Knoblauch, M. (2005). *Technologiestudie zur Verarbeitung von Polyvinylchlorid (PVC)*, commissioned by PlasticsEurope Deutschland e.V. and AGPU

Ansell, J.M. & Fowler, J.A. (1988). The acute oral toxicity and primary ocular and dermal irritation of selected N-Alkyl-2-Pyrrolidones, *Food and Chemical Toxicology*, Vol.26, No.5 (1988) p. 475

Hohenberger, W. (2001). In: *Plastics Additives Handbook*, 5th ed., Zweifel, H. ed., 901, Hanser, Munich

Third National Report on Human Exposure to Environmental Chemicals, U.S. CDC, July 2005

H.R. 4040 - 110th Congress (2007): Consumer Product Safety Improvement Act of 2008,

2

By-Products From Jumbo Squid (*Dosidicus gigas*): A New Source of Collagen Bio-Plasticizer?

Josafat Marina Ezquerra-Brauer*, Mario Hiram Uriarte-Montoya, Joe Luis Arias-Moscoso and Maribel Plascencia-Jatomea
Departamento de Investigación y Posgrado en Alimentos/Universidad de Sonora México

1. Introduction

The applications of jumbo squid by-products as plasticizer agents have been never received much attention, even though there are several studies in which the properties of the catch and processing discards for productions of films have been reported.

The principal by-products that result from the catch and processing of seafood include viscera, heads, cut-offs, bone, and skin. Therefore, a large and considerable volume of solid waste is obtained, constituting an important source of environmental contaminants unless efforts for their recovery are attained (Arvanitoyannis & Kassaveti, 2008). In the case of squid, one the largest known mollusks, the global capture represents no more than 2% of the total catch. However, by-products from squid processing, which include heads, viscera, backbones or pens, ink, skin, unclaimed fins, mantles, and tentacles, may represent up to 60% of the whole weight. In addition, from all the different anatomical squid components regarded as by-products only the beak and pen are not edible. Thus, since most of the squid is not used, its by-products also pose an environmental issue for this fishery, especially in areas where it is harvested the most. The valuable and profitable components that these by-products contain include among others chitin, chitosan, collagen, and gelatin (Kim & Mendis, 2006; Shahidi, 2006).

Collagen is a fibrous protein responsible for structural sustaining of several animal tissues, being the main protein present in skin, bones, tendons, cartilages, and teeth. In the case of mammalians, it accounts for about 20-30% from the total body protein (Quereshi et al., 2010). The term collagen derives from the Greek word *kolla* which means glue and was defined as "that constituent of connective tissue which yield gelatin on boiling" (Oxford University, 1893). Nowadays, although collagen is most of the time referred as to a single item, in fact it is a heterogeneous group of at least 19 different molecules which have a unique triple-helix configuration that forms very strong fibers. Collagen is characterized by its unusual amino acid composition, in which glycine and proline account for about 50% of them. Indeed, each polypeptide called α chain consists of a repeated sequence of the triplet Gly-X-Y, where X and Y are often proline and hydroxyproline. Each collagen type varies in the length of the helix and the nature and size of the non-helical portions (Lee, C. R. et al., 2001).

Collagen *per se* is regarded as one of the most useful biomaterials. The excellent biocompatibility and safety due to its biological characteristics, such as biodegradability and weak antigenicity, have made collagen one of the primary resources in medical applications. In addition, other uses include gelatin production, nutritional supplements, sausage casings, and in cosmetic products it claims anti-ageing benefits (Kim & Mendis, 2006; Lee et al., 2001).

Recently, the use of fish collagen in the manufacture of biopolymer films has been reported. Collagens from different species of fish have been extracted using acetic acid, which were used later to produce biodegradable films (Venugopal, 2009). Studies on the production and characterization of films using fish gelatins are quite recent, and all fish gelatins have been observed to exhibit good film-forming properties, yielding transparent, nearly colorless, water soluble, and highly extensible films (Avena-Bustillos et al., 2006; Benjakul et al., 2006; Carvalho et al., 2008; Gomez-Guillen et al., 2007; Zhang et al., 2007).

On the other hand, chitosan is a polysaccharide that is produced by deacetylation of naturally occurring chitin, and it has a great potential for a wide range of applications due to its versatile properties, such as in food and nutrition, biotechnology, material science, drugs and pharmaceuticals, agriculture and environmental protection, and recently in gen therapy as well (Venugopal, 2009; Dutta et al., 2009; Shahidi et al., 1999; Shahidi et al., 2002). Nonetheless, pure chitosan, as a film material, does not form films with adequate mechanical properties due to its low percentage of elongation (Butler et al., 1996). For this reason, one of the current trends in designing biodegradable materials for packaging is to combine different biopolymers (Bawa et al., 2003; Bertan et al., 2005; Colla et al., 2006; Le-Tien et al., 2004; Lee et al., 2004; O'Sullivan et al., 2005; Tapia-Blacido et al., 2007; Yu et al., 2006).

Chitin and chitosan belong to a group of natural polymers produced by the shells of crab, shrimp, and lobster. In addition to be nontoxic, chitin and chitosan are inexpensive, biodegradable, and biocompatible. Regarding to film-forming properties, chitosan is more versatile as compared to its precursor chitin. Chitosan has the capacity to form semipermeable coatings which, when used in foods, prolong their shelf life by acting as barriers against air and moisture (Agulló et al., 2004).

Furthermore, since collagen in acid solution exhibited positively charged groups, it has a molecular interaction with chitosan with high potential to produce biocomposites (Liang et al., 2005; Lima et al., 2006; Sionkowska et al., 2006; Wang et al., 2005; Wess et al., 2004), acting as a possible plasticizer agent.

In the first part of this chapter, the most important characteristics of jumbo squid as fishery, as well as the most recent scientific literature dealing with chitosan and collagen films made from seafood by-products, are reviewed. In the second part, thermal, mechanical and morphological properties of chitosan and acid soluble collagen (ASC) produced by casting films are discussed. As-cast films dried in relation to the molecular interaction of ASC by using differential scanning calorimetry (DSC), scanning electron microscope (SEM), and infrared spectroscopy are also discussed in this chapter. Thermal properties by DSC, SEM images, mechanical properties, water vapor barrier properties, and water solubility characteristics of the chitosan/ASC blends are analyzed as a function of ASC content in terms of the individual properties of chitosan and ASC.

2. Characteristics of jumbo squid as fishery

Squid or calamari are cephalopods which comprises a group around 300 species, being the jumbo squid or *Dosidicus gigas* one of them (Figure 1). Jumbo squid is a member of the flying squid family, Ommastrephidae (Nesis, 1985), and are known to eject themselves out of the water to avoid predators. Jumbo squid are the largest known mollusks and the most abundant of the nektonic squid. They can reach up to 2 m in length and weigh up to 45 kg. This specie is characterized by its large, tough, thick-walled mantle and long tentacles. These organisms are aggressive predators. Jumbo squid earned the nickname of "red devils" because of their red hue when hooked, which they use to camouflage from predators in deep waters where most animals cannot see the red color. This coloration is due, like other cephalopods, to the presence of chromatophores. Also, squid possess the ability to squirt ink as a defense mechanism (Nigmatullin et al., 2001).

Fig. 1. Jumbo squid (*Dosidicus gigas*).

Jumbo squid is an endemic species to the Eastern Pacific, ranging from northern California to southern Chile and to 140 degrees W at the equator. Exploratory commercial fishing for *Dosidicus gigas* began in the 1970s off the Pacific coast of America. The catches of this fishery increased from 14 tons per year in 1974 to over 250,000 tons in 2005. Since then, it has become an extremely important fisheries resource in the Gulf of California, Costa Rica Dome and Peru (Marakadi et al., 2005).

The commercial fishery of jumbo squid consists of a multinational jigging fleet, which fish at night using powerful lights to attract squid (Waluda et al., 2004). The caught of this organism depends of the season and the region. In the Gulf of California for example, this organism enter to the Gulf from the Pacific in January, to reach their northernmost limit by April, and to remain in the central Gulf from May through August; the highest aggregations of specimens are found along the western (Baja California) coast. From September squids appear to migrate onward the eastward to the Mexican mainland coast and then southwards, to the Gulf back into the Pacific (Ehrhardt et al., 1983). Whereas, in Peruvian waters the highest squid concentrations occur along the coast of northern Peru, from Puerto Pizarro to Chimbote, with low to medium squid concentrations off Pisco and Atico. The highest catches occur during autumn, winter, and spring, since squid tend to be dispersed in summer (Taipe et al., 2001).

Although the growth of this fishery has been spectacular, great contrasts have characterized it. Of the total catch, a major portion remains unused or minimally used. In Mexico for instance, no more than 11% of the resource is used for human consumption, regardless of its low price and high nutritional value (De la Cruz et al., 2007). Moreover, only the mantle

(42%) is usually used, which is later primarily marketed fresh, frozen or pre-cooked (Luna-Raya et al., 2006). Some of the by-products produced after filleting, like fins and heads, are utilized but huge amounts are wasted. Fortunately, a number of studies have reported that this waste is an excellent raw material to obtain important by-products with high commercial value, such as collagen (Gómez-Guillen et al., 2002; Kim et al., 2005, Shahidi, 2006; Torres-Arreola et al., 2008; Gimenez et al., 2009).

3. Collagen from jumbo squid by-products

Collagen is the main fibrous component of the connective tissue and the single most abundant protein in all organisms, since it represents up to 30% of the total protein in vertebrates and about 1-12% in aquatic organisms (Brinckmann, 2005). In general, collagen fibrils in the muscle of fish, form a delicate network structure with varying in the different connective tissues and is responsible for the integrity of the fillets (Shahidi, 2006). Moreover, the distribution of collagen may reflect the swimming behavior of the species (Sikorski et al., 1994). In several species of fish the weakening of the connective tissues may lead to serious quality deterioration that manifests itself by disintegration of the fillets. Also, thermal changes in collagen contributes to the desirable texture of the meat, however when heating is conducting under not controlled conditions, this may lead to serious losses due to the reduction in the breaking strength of the tissues (Sikorski et al., 1986).

The name collagen is in fact a generic term for a genetically distinct family of molecules that share a unique basic structure: three polypeptide chains coiled together to form a triple helix. About 19 different types of collagen molecules have been isolated and these not only varies in their molecular assembly, but also in their size, function, and tissue distribution (Table 1) (Exposito et al., 2002; Brinckmann, 2005).

Type	Chain composition	Subfamily	Tissue distribution
I	$[(\alpha1(I))_2\alpha2(I)]$	Fibrillar	Skin, tendon, bone, ligament, vessel
II	$[(\alpha1(II))_3]$	Fibrillar	Hyaline cartilage, vitreous
III	$[(\alpha1(III))_3]$	Fibrillar	Skin, vessel, intestine, uterus
IV	$[(\alpha1(IV))_2\alpha2(IV)]$	Network	Basement membranes
V	$[\alpha1(V)\alpha2(V)(3(V)]$	Fibrillar	Bone, skin, cornea, placenta
VI	$[\alpha1(VI)\alpha2(VI)\alpha3(VI)]$	Network	Interstitial tissue
VII	$[(\alpha1(VII))_3]$	Anchoring fibrils	Epithelial tissue
VIII	$[\alpha1(VIII)\ \alpha2(VIII)]$	Network	Endothelial tissue, descement's membrane
IX	$[\alpha1(IX)\alpha2(IX)\alpha3(IX)]$	FACIT*	Cartilage, cornea, vitreous
X	$[(\alpha1(X))3]$	Network	Hypertrophic and mineralizing cartilage
XI	$[\alpha1(XI)\alpha2(XI)\alpha3(XI)]$	Fibrillar	Cartilage, intervertebral disc
XII	$[(\alpha1(XII))_3]$	FACIT*	Skin, tendon, cartilage

* Fibril associated collagen with interrupted triple helices
Adapted from: Friess, 1998; Brinckmann, 2005.

Table 1. Chain composition, subfamily and body distribution of the main collagen types found in animal tissue.

The most abundant and widespread family of collagens, it is represented by the fibril-forming collagens, especially types I, III and V (Sato et al., 1989). Their basic structure consists of three protein chains that are supercoiled around a central axis in a right-handed manner to form a triple helix, called tropocollagen, which is a cylindrical protein of about 280 nm in length and 1.5 nm in diameter. Each chain, called α chain, contains about 1000 amino acids and it has a molecular weight of 100 kDa, depending on the source. Tropocollagen molecules may be formed by three identical chains (homotrimers) as in collagens II and III, or by two or more different chains (heterotrimers) as in collagen types I, IV, and V.

The three α chains are perfectly intertwined throughout the tropocollagen molecule to form the tripe helix except for the ends, where helical behavior is lost, since in these regions, called telopeptides, globular proteins that are involved with intermolecular crosslinking with other adjacent molecules are found (Engel & Bachinger, 2005). A structural pre-requisite for the assembly of a continuous triple helix, the most typical conformation of collagen, is that every third position along the polypeptide chain is occupied by a glycine residue. Being glycine the smallest amino acid and lacking of a side chain, the collagen α chains can coil so tightly because glycine can be easily accommodated in the middle of a steric smooth superhelix and form stable packed structures; this would be very difficult with the bulkier residues. Further stabilization of the triple helix is attained by the formation of hydrogen bonds that are formed between the amino groups of glycine residues and the carbonyl groups of residues from other chains.

The structural constraints that make the collagen triple helix unique among proteins are given by its unusual amino acid content. Besides glycine as the major residue, a repeated sequence that characterizes the collagenous domains of all collagens is the triplet (Gly-X-Y), where X and Y are often proline and hydroxiproline, respectively (Figure 2). Depending on the collagen type, specific proline and lysine residues are modified by post-translational enzymatic hydroxylation. These imino acids permit the sharp twisting of the collagen helix and are associated with the stability and thermal behavior of the triple helical conformation.

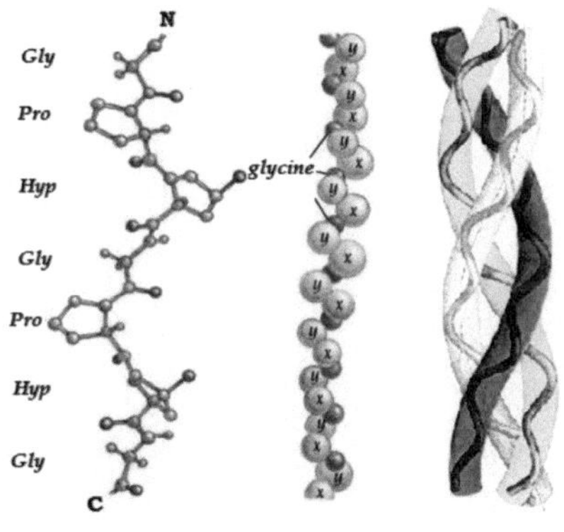

Source: Branden & Tooze, 1999; Voet & Voet, 1995.

Fig. 2. Spatial conformation of a typical tropocollagen molecule.

3.1 Sources and extraction

The main sources of industrial collagen are limited to the skin and bones of pigs and cattle. However, as a possible alternative to the problems associate to transmissible bovine spongiform encephalopathy (BSE) and foot and mouth disease (FMD), as well as religious barriers, new alternatives are being sought for collagen sources (Jongjareonraka et al., 2005; Nagai, 2004). As a result, the extraction of collagen and their derivatives from marine sources has considerably increased in the last years since it represents an appropriate alternative source to land animals with promising functional properties (Shen et al., 2007).

Collagen typically exists in a concentration from 3 to 11.1% in the mantle of some squid species like *Illex* and *Loligo* (Sikorski & Kolodziejska, 1986), whereas in *Dosidicus gigas*, collagen was found in a concentration up to 18.33% (Torres-Arreola et al., 2008). This variability among squid species may be attributed to the high degree of protein turnover that takes place in the muscle of cephalopods, which are fast-growing species as they usually reach their maximum maturity in one year or two. In general, the collagen from aquatic organisms, is highly soluble in salt solutions, dilute acids and acid buffers, unlike collagen from land mammals that are poorly soluble in such solvents (Kolodziejska et al., 1999). Consequently, in order to extract collagen from marine organisms, the most common solvent systems that have been found to be generally useful and convenient are described below; however, there is no single standard method for the isolation of collagen (Miller & Rhodes, 1982).

Prior to the actual collagen extraction procedure, it is a prerequisite to wash or digest the original tissue with a dilute alkali (*e.g.* 0.1 M NaOH) in order to remove non-collagenous proteins and to prevent the effect of endogenous proteases on collagen. This procedure can also be achieved by using caothropic solutes such as urea in high concentrations (6 M). These organic solutes are capable of increasing the ionic strength of the medium and breakdown the structure of water, which causes the disruption of the hydrogen bonds among the α chains, which induces the unfolding and solubilization of hydrophobic residues inside the protein molecule, only affecting the soluble protein fractions (myofibrilar and sarcoplasmic) but leaving the stromal proteins intact (Usha & Ramasami, 2004). Once the connective tissue have been isolated, the following collagen extraction procedures are usually conducted at low temperatures (4-8°C) with the aim to minimize bacterial growth, enhance the solubility of native collagens, and to ensure the retention of native conformation on the part of the solubilized collagens (Miller & Rhodes, 1982):

a. Neutral salt solvents (*e.g.* 1 M NaC1, 0.05 M Tris, pH 7.5): The high ionic strength of this solvent system allows to solubilize newly synthesized or young collagen molecules (Friess, 1998). Moreover, this solvent is also capable of solubilize more components beyond the stromal fractions, such as remaining myofibrillar proteins. Therefore, this solvent shows the least ability to solubilize pure collagen.
b. Dilute acid solvents (e.g. 0.5 M acetic acid): Dilute acids such as acetic acid, hydrochloric acid, or citrate buffer are widely used to dissolve some pure collagen, although they are only limited to the portion of non-crosslinked collagen (Jongjareonrak et al., 2005). The pH of these solvents is about 3.0, which exhibits a sufficient capacity to induce swelling of most tissues promoting the solubilization of the triple helix junctions. Collagen extracted using these solvents usually have greater industrial applications in comparison with other soluble fractions because of the facility to incorporate the biomolecule with other polymers in an acidic medium (Garcia et al., 2007).
c. Dilute acid solvents containing pepsin (*e.g.* 0.5 M acetic acid plus EC 3.4.23.1): These systems are one of the most versatile and widely used procedures for the extraction of collagen. The enzyme is usually added to the acidic solvent in sufficient quantities to

achieve a 1:10 ratio between the enzyme and the dry weight of the tissue to be extracted. The effectiveness of this system lies in its ability to solubilize native collagen (Miller & Rhodes, 1982). Despite its great efficacy, an essential and adequate control is needed for these systems, since pepsin is an exogenous proteolytic enzyme and if not controlled, it may contribute to the total disorganization or degradation of the collagen molecules.

After the different soluble collagen fractions are attained, the insoluble collagen remains. This fraction is characterized by numerous crosslinks within the molecule, which in turn prevent its disintegration and make it more resistant to proteolysis. Physically, insoluble collagen is an opalescent fibrous material that can only be degraded using either strong denaturing agents or mechanical fragmentation under acidic conditions (Friess, 1998). The insolubility of collagen is attributed to the modification of the forces that hold in together the α-chains. Its presence is also related to the age of the animal and to its frozen storage, due to the protein aggregation that results from the removal of water molecules from the stromal proteins (Montero et al., 2000). Finally, once all the soluble and insoluble collagen fractions are collected, a selective neutral or dilute acid- salt precipitation procedure is usually performed in order to initially purify and recover the individual collagen types that may be present in extracts from various tissues. In the case that a final purification and resolution of the collagen molecules is required, different chromatographic techniques under non-denaturing conditions have been reported and found to be quite effective.

In the case of the mantle of *Dosidicus gigas*, Uriarte-Montoya et al. (2010) used urea 6M and the three solvent systems described above to isolate collagen from this species. They found that collagen from the mantle was 37% soluble in a neutral salt solvent, 23% in a dilute acid solvent and 25% in a dilute acid solvent containing pepsin, whereas the remaining 15% of the connective tissue was insoluble. They concluded that depending on the ultimate application, each collagen fraction warrants sufficient importance and might be used for different purposes rather than the traditional industrial collagen applications. As it will be mentioned in the next subsection, collagen has a wide variety of application, from food to medical and it is widely use in the form of collagen casings.

3.2 Collagen as biomaterial and applications

Nowadays, collagen has several industrial and biomedical applications. In the former, collagen has been used from long time ago, whereas the latter have placed collagen as an object of intense research in the last years. Industrially, collagen has been used for leather processing and gelatin production. Both products consist mainly of collagen but they greatly differ in the chemical form of the collagen used (Meena et al., 1999). Leather is basically chemically treated animal skin, while gelatin is an animal connective tissue that is denatured and degraded by heat and chemicals to produce a soluble form.

Regarding the biomedical uses of collagen, probably one of the main attraction of collagen as a biomaterial is its low immunogenicity. Moreover, collagen can be processed into various presentations, such as sheets, tubes, sponges, powders, injectable solutions and dispersions, making it a functional component for specific applications in ophthalmology, wounds and burns, tumor treatment, engineering and tissue regeneration, among others areas (Kim & Mendis, 2006). Table 2 summarizes the main advantages and disadvantages of collagen as a biomaterial. Another recent application of collagen in biomedicine is found in the formulation of membranes and hydrogels, products in which collagen interacts with another material to form a composite. The combination of collagen with chitosan is perhaps one of the most studied composites with practical emphasis on medicine, dentistry, and pharmacology (Lima et al., 2006).

Advantages	Disadvantages
Available in abundance and easily purified from living organisms	High cost of pure type collagen
Non-antigenic and non-toxic	Variability of isolated collagen
Biodegradable and biocompatible	Hydrophilicity may lead to swelling and more rapid release
Synergic with bioactive compounds	Variability in enzymatic degradation rate as compared with hydrolytic degradation
Formulated in a number of different presentations	Complex handling properties
Hemostatic	Possible side effects such as mineralization
Easily modifiable to produce other materials	
Plasticity due to high tensile strength	
Compatible with synthetic polymers	
Adapted from: Friess, 1998; Lee et al., 2001.	

Table 2. Main Advantages and disadvantages of collagen as a biomaterial.

4. Chitosan

Chitosan, the main deacetylated derivative of chitin, is a biocompatible, non-toxic, edible, biodegradable polymer, and it possesses antimicrobial activity against bacteria, yeast and fungi, including toxigenic fungi (Cota-Arriola et al., 2011). This polymer is the only natural cationic polysaccharide, which special characteristics makes it useful in numerous applications and areas, such medicine, cosmetics, food packing, food additives, water treatment, antifungal agent, among others (Hu et al., 2009). At a commercial level, chitosan is obtained from the thermo-alkaline deacetylation of chitin from crustaceans (Kurita, 2006), and its production has taken great importance in the ecological and economic aspects, due to the use of marine by-products (Nogueira et al., 2005).

Chemically, chitosan is a linear polycationic hetero-polysaccharide (poly [β-(1,4)-2-amino-2-deoxi-D-glucopyranose]) (Figure 3) of high molecular weight, whose polymeric chain changes in size and deacetylation degree (Tharanathan & Srinivasa, 2007). The nitrogen forms a primary amine and causes N-acylation reactions and Schiff alkalis formation, while the hydroxyl (OH) and amino (NH_2) groups allow the formation of hydrogen bonds (Agulló et al., 2004).

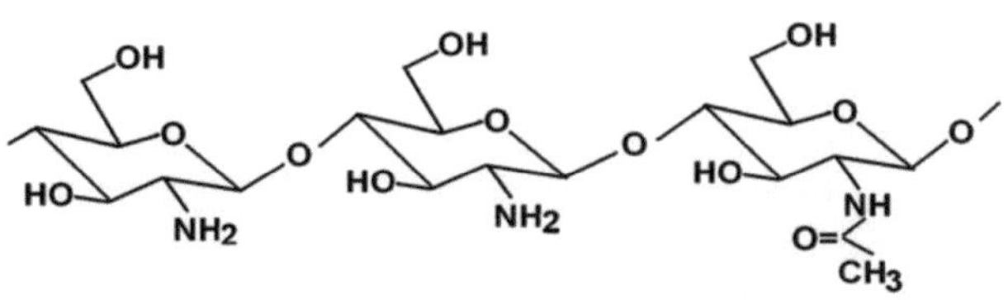

Fig. 3. Chemical structure of chitosan showing the characteristic amino (NH_2) groups.

In its native form, the chitosan is insoluble in water and in the majority of organic solvents; nevertheless, at pH values <6.0 chitosan easily dissolve in diluted acidic solutions due to the protonation of the amino groups, turning it into a soluble cationic polysaccharide (Agulló et al., 2004; Raafat & Sahl, 2009; Sharma et al., 2009). Commercially, chitosan is available in several grades of purity, molecular weight, distribution and length of chain, deacetylation degree, density, viscosity, solubility, water retention capacity and the distribution of the amino/acetamide groups. All these characteristics affect the physicochemical properties and therefore, its application (Raafat & Sahl, 2009).

Among their biological properties, the antibacterial, antifungal, antiviral and insecticide activity of chitosan and its derivatives (Badawy & Rabea, 2005; Badawy et al., 2005) have been explored for agricultural applications (Daayf et al., 2010). Due to its peculiar characteristics, chitosan has potential applications in diverse fields (Table 3).

Industry	**Applications**
Biomedicine	It promotes the growth of tissues, wound healing (bandages, sutures), flocculent, homeostatic, bacteriostatic/fungistatic, spermicide, antitumoral, anticholesterolemic, immunoadjuvant, sedative of the central nervous system, acceleration of the osteoblast's formation.
Pharmaceutics	Controlled release of medicines.
Agriculture	Preservation of fruits and vegetables, soils supplements, release of agrochemicals and nutrients supply, improves the seed's germination, protects against microbial damage.
Waste-water treatment	Bleaching of waste water, removal of heavy metals, water clarification.
Paper and textile	Flexibility and resistance. It improves stickiness of inks and clothes stains, stabilizes the color and resistance of cloths. It improves the paper sheen, the resistance to the microbial and enzymatic deterioration; improves the paper biodegradability and the antistatic of photographic paper, improves the paper air-tightness.
Cosmetics	Solar and moisturizing protector, decrease the expression lines, useful in contact glasses and hair conditioners.
Chromatography	Enzymes separation and gas chromatography.
Supplement or food additive	Water retention, reduces the lipids and cholesterol absorption, antiacid agent, food fibre, food formulation for babies, emulsifier, preservative, antioxidant, gellificant, clarficant of fruit juices, immobilization of enzymes.
Films and eatable recovering	It protects and preserves food, decreases the production of ethylene and CO_2, antimicrobial, reduces the water permeability, controls enzymatic oxidation.
Preservative	Antibacterial, antifungal.

Adapted from: Agulló et al., 2004

Table 3. Applications of chitosan and chitosan composite films.

4.1 Chitosan bio-based films

One of the properties of major importance of chitosan is its filmogenic capacity. The films can be prepared from moderately concentrated polymer solutions, in general at 3 % (w/w). Chitosan films possess acceptable mechanical and permeability properties, good adherence to different surfaces, are flexible, resistant to water and show excellent barrier properties against gases (O_2, CO_2, water vapor), that allows its application in the development of food packing materials (Agulló et al., 2004; Plascencia-Jatomea et al., 2010; Tharanathan et al., 2002).

There are different methods for the productions of chitosan films, as the evaporation of solvents (Casting) -the first developed and more used at present- and the extrusion with some polyesters, olefins or carbohydrates. The last one is the most used for the industrial production of polymeric materials (Bhattacharya et al., 2005; Pelissari et al., 2009). With regard to the edible covered films, these are obtained by direct application (spraying, immersion) of chitosan solutions on the food surface or in the medicine's tablets, forming a thin layer that covers and protects the product of the environment (Janjarasskul & Krochta, 2010). Nowadays, most of the films and chitosan composites are prepared by this method, changing the solvent and the component's concentration of the blends, according to the application. Recently, it has been reported that the electrospinning technique allows the preparation of ultrathin chitosan nanofibers with unusually high porosity in their nanometer scale architecture and large surface area (Chen, Z. et al., 2009; Martínez-Camacho et al., 2011; Ohkawa et al., 2004). As particular interests have been addressed in the tissue engineering, great efforts have been made to study electrospinning of biodegradable polymers (Schauer & Schiffman, 2008).

Due to its abundance in the nature and to its biocompatibility, chitosan is considered to be a promising polymer for the development of functional materials (Ohkawa et al., 2004; Westbroek et al., 2007). In contrast to other materials, it has been demonstrated that chitosan films possess antifungal properties (Martínez-Camacho et al., 2011; Plascencia-Jatomea et al., 2010), that make it a good alternative for food protection and food shelf life extension (Chien et al., 2007b; Chien et al., 2007a; Coma et al., 2003; El Ghaouth et al., 1992; Fornes et al., 2005; Li & Yu, 2001; No et al., 2007; Schnepf et al., 2000). In general terms, although the materials prepared with conventional synthetic polymers are functional, of easy production, and low cost, they hardly are degradable, which strongly impacts the environment (Tharanathan & Srinivasa, 2007). Nevertheless, the use of these biopolymers is limited due to problems related to its deficient mechanical properties (fragility, poor barriers against gases and moisture) and cost (De Azeredo, 2009).

The incorporation of natural compounds has allowed the appearance of new materials with good mechanical properties, which overcome those that possess the individual materials. Most of these films are prepared mainly thinking about its use as food packing and tissue engineering materials (Tharanathan & Srinivasa, 2007). Additionally, to improve chitosan blends elasticity a biodegradable materials might be added (Butler et al., 1996).

4.2 Chitosan films and bio-plasticizer

Plasticizers are additives used to increase the flexibility or plasticity of polymers (Daniels, 1989). The most studied plasticizing agent in chitosan films has been glycerol and polyols and its efficacy in improving the properties of films has been well-documented (Table 4).

Film/Plasticizer	Physicochemical properties	Reference
Chitosan Lignine	It improves the tensile force, the thermal degradation and the glass transition temperatures.	(Chen et al., 2009)
Starch-chitosan film/ Sorbitol; glycerol; polyethylene glycol	Increased flexibility and barrier properties; decreased rigidity; increased in solubility; affected color.	(Bourtoom, 2007)
Chitosan film/ Glycerol; ethylene glicol; poly (ethylene glycol); propylene glycol	Improves ductility, increase in strain and decrease in stress, and increases film hydrophilicity. Propylene glycol exposes antiplasticization phenomenon.	(Suyatama et al., 2005)
Chitosan film/ Glycerol; sorbitol; i-erythritol	Good flexibility and mechanical force.	(Fernández et al., 2004)
Chitosan film/ Poly (ethylene glycol)	Improve the elastic properties, enhanced the protein adsorption, cell adhesion, growth and proliferation.	(Zhang et al., 2002)
Chitosan film/ Poly(vinyl alcohol; sorbitol; sucrose	Decrease in thermal properties, increase in percentage of elongation and CO2 and water permeability increase. High plasticizer contents caused decrease in both tensile and modulus.	(Arvanitoyannis et al., 1997)
Chitosan films/ Glycerol	Tensile strength values equal to that of HDPE and LDPE films	(Butler et al., 1996).

Table 4. Plasticizer effect on the properties of chitosan films.

Recently, the potential action of acid soluble collagen from jumbo squid (ASC) as a plasticizer in chitosan films was evaluated (Uriarte-Montoya et al., 2010; Arias-Moscoso et al., 2011). In general, the use of acid soluble collagen from jumbo squid by-products, in amounts equal or lower than 50 % in the production of chitosan biocomposites, produces films with reasonable tensile properties (Arias-Moscoso et al., 2011). More details about it will discuss in the following subsections.

5. Chitosan and collagen films

Biomaterials are increasingly being used in several fields, like packaging material (Tharanathan, 2003), tissue engineering (Nalwa, 2005), medical and pharmaceutical applications (Bures et al., 2001) due to their functional properties.

Collagen, as a natural polymer, has very weak mechanical properties, especially in aqueous media (Zhang et al., 1997). On the other hand, as it was mentioned previously, chitosan poses excellent film-forming property, antimicrobial activity, and unique coagulating ability with metal and other lipid and protein complexes due to the presence of a high density of amino groups and hydroxyl groups in the polymeric structure of chitosan (Dutta et al., 2002; Li et al., 1992; Shahidi et al., 1999). Although, mechanical properties of chitosan films are comparable to those of many medium-strength commercial polymers, it is considered that this properties could be improved, mainly its elasticity (Suyatma et al., 2005). One strategy to increase chitosan elasticity is to associate it with biodegradable materials. However, the interactions that may occur between biopolymers are very important when the characteristics of any material are considered to be transformed. These interactions depend on the miscibility of its components.

Miscibility in polymer blends is attributed to specific interactions between polymeric components, which usually give rise to a negative free energy of mixing in spite of the high molecular weight of polymers (Shanmugasundaram et al., 2001). On the other hand, the interactions between natural polymers of different chemical structures, whether they are hydrogen bonding or electrostatic in nature, considerably improve the mechanical properties of the material obtained from such mixtures (Zhang et al., 1997). Although, most of polymers blends are immiscible with each other due to the absence of specific interactions, TEM micrographs of collagen–chitosan composites have shown that chitosan network can interpenetrate into the collagen network; the chitosan phase is wrapped in the collagen phase and is denser; besides, the amount of chitosan phase grows with the content of chitosan increasing (Zhang et al., 1997).

The main kinds of interactions that can give rise between the two polymers when they are in contact with water are: an electrostatic complex and an hydrogen bonding type of complex, in the presence of a great excess of chitosan (Taravel & Domard, 1993; Taravel & Domard, 1995). Hydrogen bonds between collagen and chitosan can be formed as follows: between either a carbonyl, hydroxyl, or an amino group from collagen, and either hydroxyl, amino, or a carbonyl group from chitosan. The formation of hydrogen bonds between two different macromolecules competes with the formation of hydrogen bonds between molecules of the same polymer (Mo et al., 2008). Apparently, interactions between collagen and chitosan depend on the structural organization of collagen and the amount and distribution of charges along the polymer chains. These properties are directly related to the pH of the medium, which is of fundamental importance for the study of interactions present between the biopolymers (Tohni, 2002).

5.1 Stability

It is known that some environmental conditions may induce degradation of biopolymers, mainly affecting the mechanical properties and limiting their presentation, therefore the stability of the film is also a very important characteristic to evaluate. Lima et al. (2006) detected that a more stable film can be obtained when the chitosan in a collagen-chitosan film reaches 50% because chitosan increases the organization of the microscopic structure of the collagen. The effect of solar radiation on the properties of collagen, chitosan, and collagen-chitosan films, have been reported. Research of Sionkowska (2006), Sionkowska et al. (2006), and Sionkowska et al. (2011) indicated that collagen-chitosan blends are more sensitive to the action of UV irradiation or artificial solar light than pure collagen or pure chitosan films.

5.2 Applications of collagen/chitosan blends

Biomaterials from collagen/chitosan blends with different applications have been successfully obtained (Table 5). However, it is important mentioned that the biological or mechanical properties, as well as stability of the product, depends of the compositions of the blend (Table 5).

Composition	Main characteristics	Reference
Collagen/Chitosan composite hydrogel	Hydrogel system stable Capable of encapsulating Tβ4	(Chiu & Radisic, 2011)
Gelatin -Chitosan films	Water resistant Deformable Increase in breaking strength Rubbery semi-crystalline materials Antimicrobial activity Biodegradable films	(Gómez-Estaca et al., 2011)
Collagen/Chitosan scaffolds fabricated via thermally triggered cofibrillogenesis approach	Electrostatic interactions Intereconnected porous structure Excellent mechanical properties Collagen-chitosan scaffolds with superior characteristics	(Wang & Stegemann, 2011)
Electrospun 80 % Collagen/20 % Chitosan fibers	Fibrous membrane softer, flexible and elastics	(Chen, Z. et al., 2009)
Electrospun 20 % Collagen/80 % Chitosan fibers	Fibrous membrane inflexible, less compact	(Chen et al., 2009)
Anionic collagen /Chitosan sponges by freeze lyophilization	Polyelectrolytic interaction Low thermal stability Small porous size Promising feature to be processed into porous structures Application in cell transplantation and tissue regeneration	(Horn et al., 2009)
Chitosan/Gelatin copolymer	Very suitable for coating meat products	(Shane & Champa, 2009)
Collagen/Chitosan Scaffold	It accelerates of cellular proliferation; It reduces the biodegradation rate. Materials for tissue engineering.	(Tangsadthakun et al., 2007)
70% Collagen-30% Chitosan Films	Thermally stable Less mechanical stable	(Sionkowska et al., 2006)
50% Collagen- 50% Chitosan Films	Low thermal stability Low mechanical stability	(Sionkowska, 2006)
Collagen/chitosan composite microgranules	Effective for the controlled release of transforming growth factor beta 1 Satisfy specifications for cartilage tissue engineering	(Lee et al., 2006)
Collagen /Chitosan Scaffold	Porous composite matrix Good tensile strength Biocompatible and biodegradable May be used as a chondrocyte carrier for cartilage tissue engineering	(Shi et al., 2005)
Complex membranes of collagen/chitosan.sodium hyaluronate	Material suitable for tissue-engineered cornea	(Chen et al., 2005)

Table 5. Properties of collagen/gelatin-chitosan blends.

6. Chitosan and acid soluble collagen form jumbo squid by-products blend films

It has been demonstrated that collagen/chitosan is miscible at any composition range in acetic acid solution and the compatibility will remain even when the solvent is absent, being possible the compatibility in a solid state (Dan et al., 2007). Collagen in acid solution exhibited positively charged groups and it has a molecular interaction with chitosan (Figure 4) with high potential to produce biocomposites with novel properties as was previously described (Liang et al., 2005; Lima et al., 2006; Sionkowska et al., 2006; Wang et al., 2005; Wess et al., 2004). Besides that, acid soluble collagen might be useful as a new source of plasticizer agent in the preparation of biofilms in composites with chitosan (Uriarte-Montoya et al., 2010).

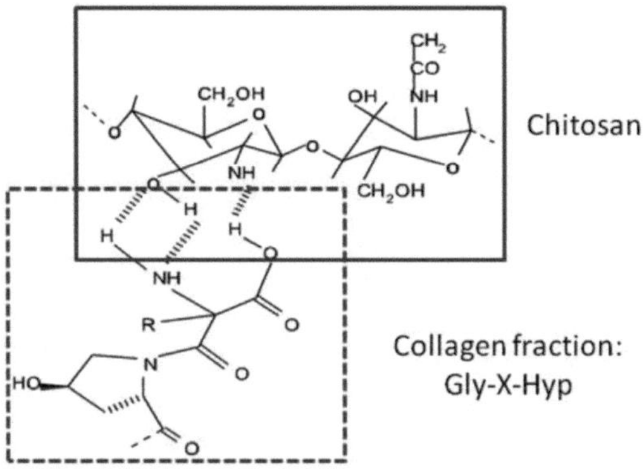

Fig. 4. Schematic presentation of chitosan-collagen fraction molecular interaction.

6.1 Molecular interactions in chitosan and acid soluble collagen from jumbo squid films

The molecular interaction between acid soluble collagen form jumbo squid (ASC) and chitosan was reported to be mainly due to hydrogen bonding between the two polymers (Uriarte-Montoya et al., 2010). These observations were based on the modification of the amplitude bands in infrared (FTIR) analysis. FTIR spectroscopy is an useful tool to study the hydrogen bonding and other interactions as well as the miscibility of polymer blends (Zhang et al., 2002).

The typical peaks of chitosan are around 2890, 1645, 1563, and 1414 cm^{-1} which correspond to aliphatic groups (-CH_2 and -CH_3), amides I, II, and vibrations of -OH groups from primary alcohols, respectively. The amide I arises from C=O stretching; the amide II arises from -NH torsion groups. When a shoulder at 1645 cm–1 is detected, it suggests that chitosan comes from a partial desacetylation process (Wess et al., 2004). The addition of ASC to commercial chitosan induced a decrease in the bands around 1645, 1563, and 1414 cm-1 (Uriarte-Montoya et al., 2010) indicating that there is an interaction between the polymers and the compatibility between collagen and chitosan, which results in good film homogeneity.

On the other hand, the differential scanning calorimetry (DSC) analysis has been used by different authors to elucidate how the collagen associates to other macromolecules (Privalov & Tiktopoulo, 1970). The glass transitions temperatures (Tg) is an important criteria for the miscibility of the components. In a completely miscible blend of two polymers, only one Tg will appear in the DSC thermograms (Suyatma et al., 2005). Uriarte-Montoya et al. (2010) and Arias-Moscoso et al. (2011) detected that in chitosan-ASC films at 85-15 and 50-50 respectively, the Tg value of was lower than that of the chitosan film. According with the theory of plasticization, those results confirm that chitosan-ASC have good miscibility (Suyatma et al., 2005), and suggests that collagen may act as plasticizer in chitosan film structure.

6.2 Characterization of chitosan/acid soluble collagen from jumbo squid blends

Films produced by mixing collagen from jumbo squid and chitosan are easily obtained by using a casting plate (Arias-Moscoso et al., 2011). These films usually are opaque, soft, with porous structure, and hygroscopic (Figure 5), which also poses poor water barrier properties and slight acidic smell (Arias-Moscoso et al., 2011). The last mentioned properties may produce films not suitable to prevent deterioration in some kinds of products; however, might be suitable for medical or pharmaceutical applications and, to avoid the acidic smell, a neutralization process may be applied.

A

B

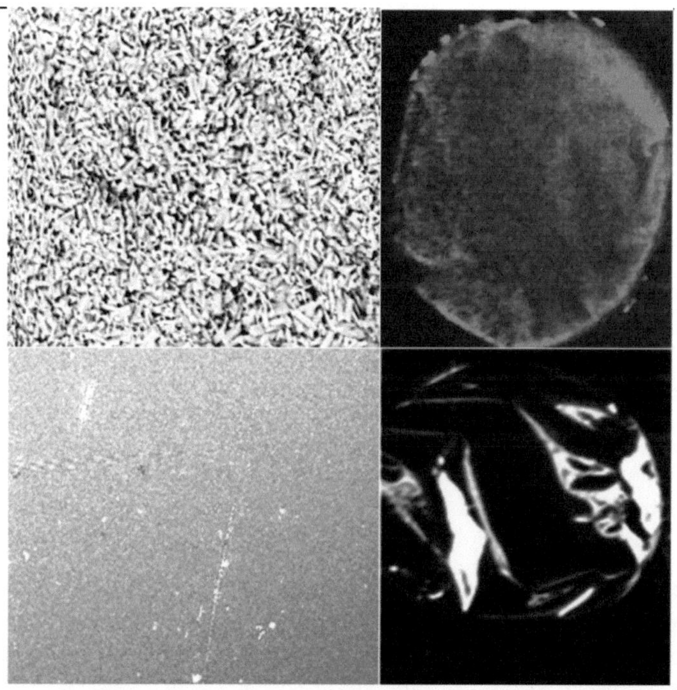

Fig. 5. Micrograph and picture of acid soluble collagen obtained from jumbo squid by-products-chitosan (A) and chitosan (B) films.

6.2.1 Mechanical properties

The acid soluble collagen from jumbo squid induces modification of crosslinking reaction in chitosan films affecting their mechanical properties (Table 6). In general, the presence of acid soluble collagen from jumbo squid on chitosan films induces the formation of more flexible and less rigid films (Arias-Moscoso et al., 2011; Uriarte-Montoya et al., 2010).

According with Arias-Moscoso et al., (2011) the general trends of the stress-strain curves of acid soluble collagen-chitosan films are characteristic of elastic films. However, these tensile properties vary with the collagen content on the film. It was detected that the tensile strength values of most of ASC/chitosan films evaluated were within the acceptance range of low density polyethylene food packaging (4 and 78.6 mPa42). Meanwhile, the elongation at the break of chitosan films with 20% of ASC was comparable to cellophane and also comparable to polypropylene when contained 50% of ASC, which for some specific applications is a desirable attribute.

Properties	Films	
	CH	ASC/CH
Tensile strength (MPa)	46.5-53.7	1.4-39.6
Elongation at break (%)	4.9-5.5	11.0-4.3
Elastic modulus (Mpa)	2210-2514	660-2430

Adapted from Arias-Moscoso et al. (2011)

Table 6. Mechanical properties of acid soluble collagen from jumbo squid-chitosan (ASC/CH) and chitosan (CH) films.

It is known that the mechanical properties are dependent on the distribution and intensity of inter- and intramolecular interaction; therefore, when the content of ASC increases, the collagen molecules also increases, and this may reduce chitosan interaction, inducing high possibilities of movements, producing films more flexible and less rigid.

7. Conclusion

It is possible to recover and utilize compounds from squid wastes, which may represent a potential for increased business and more ways to make a more environmentally sounded use of natural resources from this seafood product. From this wastes or by-products collagen can be obtained.

ACS can be extracted from the muscle jumbo squid with 0.5 M acetic acid with an average yield of 15% from the total muscle protein, and showed comparable biochemical characteristics to the collagen of skin from squid species, but it presented higher transition temperature, similar to type I collagen from bovine skin.

Novel biomaterials from collagen-chitosan blends, with excellent biocompatibility and antibacterial properties can be successfully prepared. One of the applications of these blends is the elaboration of films. Moreover, the interactions between collagen and chitosan may exhibit great potential in areas not much exploited such as plasticizers.

Chitosan, a polysaccharide that is produced by deacetylation of naturally occurring chitin, has a great potential for a wide range of applications due to its biodegradability, biocompatibility, antimicrobial activity, non-toxicity, and versatile chemical and physical properties. Acid soluble collagen/chitosan blends are miscible and interact at the molecular level, although pure chitosan, as a film material, poses low percentage of elongation, it is possible to improve its elasticity with the addition of biodegradable materials. The film properties of the chitosan can be enhanced by adding collagen to the blend, where both polymers interact by electrostatic and hydrogen bonding.

ASC from jumbo squid and chitosan blends are miscible and interact at the molecular level, being hydrogen bonding the most abundant interaction forces between the polymers. These interactions affect the chitosan films properties. Opaque and more elastic chitosan films are obtained by the incorporation of acid soluble collagen.

The general trend of the stress-strain curves of acid soluble-chitosan films was characteristic of elastic films where collagen showing a conventional action of plasticizers (increase in elongation and decrease in strength). Therefore acid soluble collagen from jumbo squid is suitable to be used as additive to improve the elastic properties of the chitosan films.

This chapter only covers a limited introduction to collagen as bio-plasticizer. However, in view of the advances in technologies for recovering collagen from seafood catch and processing discards that should environment friendly and the continue need within the seafood industry to find alternative products, further research efforts should also be directed toward ways to evaluated the industrial potential of collagen as an alternative to traditional petroleum-based plasticizer, and provide its comparison as a chitosan or other polymer plasticizer with other bio-plasticizer.

8. Acknowledgments

The authors thank Dr. Armando Burgos-Hernández and LGA Giselle Moreno-Ezquerra for their technical assistance. This study was supported by CONACYT.

9. References

Agulló, E.; Albertengo, L.; Pastor de Abram, A.; Rodríguez, S. & Valenzuela, F. (2004). *Quitina y Quitosano: Obtención, Caracterización y Aplicaciones* Pontificia Universidad Católica del Perú, Lima, Perú.

Arias-Moscoso, J. L.; Soto-Valdez, H.; Plascencia-Jatomea, M.; Vidal-Quintanar, R.-L.; Rouzaud-Sandez, O. & Ezquerra-Brauer, J. M. (2011). Composites of chitosan with acid-soluble collagen from jumbo squid (*Dosidicus gigas*) by-products. *Polymer Internacional*, Vol.60, No.6, (March 2011), pp. (924–931), 0959-8103.

Arvanitoyannis, I. S. & Kassaveti, A. (2008). Fish industry waste: treatments, environmental impacts, current and potential uses. *International Journal of Food Science and Technology*, Vol.43, No.4, (April 2008), pp. (726-745), 0950-5423.

Arvanitoyannis, I., Kolokuris, I., Nakayama, A., Yamamoto, N. & Aiba, S. (1997). Physico-chemical studies of chitosan-poly(vinyl alcohol) blends plasticized with sorbitol and sucrose. *Carbohydrate Polymers*, Vol.34, No. 1-2, (December), pp. (9-19), 0144-8617.

Avena-Bustillos, R. J.; Olsen, C. W.; Olson, D. A.; Chiou, B.; Yee, E.; Bechtel, P. J. & McHugh, T. H. (2006). Water vapor permeability of mammalian and fish gelatin films. *Journal of Food Science*, Vol.71, No.4, (May 2006), pp. (E202-E207), 0022-1147.

Badawy, M. E. & Rabea, E. I. (2005). Synthesis of some N-benzylphosphoryl chitosan derivatives and their fungicidal and insecticidal activity. *Journal of Pest Control and Environmental Sciences*, Vol.13, No.2, (January 2005), pp. (43-56), 1110-7308

Badawy, M. E. I.; Rabea, E. I.; Rogge, T. M.; Stevens, C. V.; Steurbaut, W.; Hofte, M. & Smagghe, G. (2005). Fungicidal and insecticidal activity of O-acyl chitosan derivatives. *Polymer Bulletin*, Vol.54, No.4-5, (July 2005), pp. (279-289), 0170-0839.

Bawa, A. S.; Jagannath, J. H.; Nanjappa, C. & Das Gupta, D. K. (2003). Mechanical and barrier properties of edible starch-protein-based films. *Journal of Applied Polymer Science*, Vol.88, No.1, (April 2003), pp. (64-71), 0021-8995.

Benjakul, S.; Jongjareonrak, A.; Visessanguan, W. & Tanaka, M. (2006). Effects of plasticizers on the properties of edible films from skin gelatin of bigeye snapper and brownstripe red snapper. *European Food Research and Technology*, Vol.222, No.3-4, (February 2006), pp. (229-235), 1438-2377.

Bertan, L. C.; Tanada-Palmu, P. S.; Siani, A. C. & Grosso, C. R. F. (2005). Effect of fatty acids and "Brasilian elemi" on composite films based on gelatine. *Food Hydrocolloids*, Vol.19, (June 2005), pp. (73-82), 0268-005X.

Bhattacharya, M.; Correlo, V. M.; Boesel, L. F.; Mano, J. F.; Neves, N. M. & Reis, R. L. (2005). Properties of melt processed chitosan and aliphatic polyester blends. *Materials Science and Engineering a-Structural Materials Properties Microstructure and Processing*, Vol.403, No.1-2, (August 2005), pp. (57-68), 0921-5093.

Branden, C. & Tooze, J. (1999). *Introduction to Protein Structure* (2nd), Garland Publishing, 978-0815323051, New York, USA.

Brinckmann, J. (2005). Collagens at a glance. *Collagen*, Vol.247, (June 2005), pp. (1-6), 0340-1022.

Bourtoom, T. (2008). Plasticizer effect on the properties of biodegradable blend film from rice-starch-chitosan. Sonklanakarin Journal of Science and Technology, Vol. 30, No. 1, (April, 2008), 149-165, 0125-3395.

Bures, P.; Huang, Y.; Oral, E. & Peppas, N. A. (2001). Surface modifications and molecular imprinting of polymers in medical and pharmaceutical applications. *Journal of Controlled Release*, Vol.72, No.1-3, (May 2001), pp. (25-33), 0168-3659.

Butler, B. L.; Vergano, P. J.; Testin, R. F.; Bunn, J. M. & Wiles, J. L. (1996). Mechanical and barrier properties of edible chitosan films as affected by composition and storage. *Journal of Food Science*, Vol.61, No.5, (October 1996), pp. (953-956), 0022-1147.

Carvalho, R. A.; Grosso, C. R. F. & Sobral, P. J. A. (2008). Effect of chemical treatment on the mechanical properties, water vapour permeability and sorption isotherms of gelatin-based films. *Packaging Technology and Science*, Vol.21, No.3, (May 2008), pp. (165-169), 0894-3214.

Chen, J.; Li, Q.; Xu, J.; Huang, Y.; Ding, Y.; Deng, H.; Zhao, S. & Che, R. (2005). Study on Biocompatibility of Complexes of Collagen-Chitosan-Sodium Hyaluronate and Cornea. *Journal of Artificial Organs*, Vol.29, No.2, (February 2005), pp. (104-113), 1434-7229.

Chen, L.; Fu, Q.; Tang, C. Y.; Ning, N. Y.; Wang, C. Y. & Zhang, Q. (2009). Preparation and Properties of Chitosan/Lignin Composite Films. *Chinese Journal of Polymer Science*, Vol.27, No.5, (September 2009), pp. (739-746), 0256-7679.

Chen, Z.; Wei, B.; Mo, X.; Lim, C. T.; Ramakrishna, S. & Cui, F. (2009). Mechanical properties of electrospun collagen-chitosan complex single fibers and membrane *Materials Science and Engineering: C*, Vol.29, No.8, (July 2009), pp. (2428-2435), 0921-5093.

Chien, P. J.; Sheu, F. & Yang, F. H. (2007a). Effects of edible chitosan coating on quality and shelf life of sliced mango fruit. *Journal of Food Engineering*, Vol.78, No.1, (January 2007a), pp. (225-229), 0260-8774.

Chien, P. J.; Sheu, F. & Lin, H. R. (2007b). Quality assessment of low molecular weight chitosan coating on sliced red pitayas. *Journal of Food Engineering*, Vol.79, No.2, (March 2007b), pp. (736-740), 0260-8774.

Chiu, L. L. Y. & Radisic, M. (2011). Controlled release of thymosin β4 using collagen-chitosan composite hydrogels promotes epicardial cell migration and angiogenesis. *Journal of Controlled Release*, (May 2011), 10.1016, 0168-3659.

Colla, E.; Sobral, P. J. d. A. & Menegalli, F. C. (2006). *Amaranthus cruentus* Flour Edible Films: Influence of Stearic Acid Addition Plasticizer Concentration, and Emulsion Stirring Speed on Water Vapor Permeability and Mechanical Properties. *Journal of Agricultural and Food Chemistry*, Vol.54, No.18, (August 2006), pp. (6645-6653), 0021-8561.

Coma, V.; Deschamps, A. & Martial-Gros, A. (2003). Bioactive packaging materials from edible chitosan polymer - Antimicrobial activity assessment on dairy-related contaminants. *Journal of Food Science*, Vol.68, No.9, (December 2003), pp. (2788-2792), 0022-1147.

Cota-Arriola, O.; Cortez-Rocha, M. O.; Rosas-Burgos, E. C.; Burgos-Hernández, A.; López-Franco, Y. L. & Plascencia-Jatomea, M. (2011). Antifungal effect of chitosan on the growth of Aspergillus parasiticus and production of aflatoxin B1. *Polymer International*, Vol.60, No.6, (June 2011), pp. (937-944), 0959-8103.

Daayf, F.; El Hadrami, A.; Adam, L. R. & El Hadrami, I. (2010). Chitosan in Plant Protection. *Marine Drugs*, Vol.8, No.4, (April 2010), pp. (968-987), 1660-3397.

Dan, W. H.; Ye, Y. C.; Zeng, R.; Lin, H.; Dan, N. H.; Guan, L. B. & Mi, Z. J. (2007). Miscibility studies on the blends of collagen/chitosan by dilute solution viscometry. *European Polymer Journal*, Vol.43, No.5, (May 2007), pp. (2066-2071), 0014-3057.

Daniels, C.A. (1989). *Polymers: Structure and Properties*, (First edition) Technomic Publishing, ISBN 0-87762-552-2, Lancaster, PA.

De Azeredo, H. M. C. (2009). Nanocomposites for food packaging applications. *Food Research International*, Vol.42, No.9, (Nov 2009), pp. (1240-1253), 0963-9969.

De la Cruz González, F.; Aragón Noriega, A.; Urciaga García, J. I.; Salinas Zavala, C.; Cisneros Mata, M. A. & Beltrán Morales, L. F. (2007). Análisis socioeconómico de las pesquerías de camarón y calamar gigante en el noroeste de México. *Interciencia*, Vol.32, No.3, (January 2007), pp. (144–150), 0378-1844.

Dutta, P. K.; Ravikumar, M. N. V. & Dutta, J. (2002). Chitin and chitosan for versatile applications. *Journal of Macromolecular Science, Part C: Polymer Reviews*, Vol.42, No.3, (January 2002), pp. (307-354), 0022-2348.

Dutta, P. K.; Tripathi, S.; Mehrotra, G. K. & Dutta, J. (2009). Perspectives for chitosan based antimicrobial films in food applications. *Food Chemistry*, Vol.114, No.4, (June 2009), pp. (1173-1182), 0308-8146.

Ehrhardt, N. M.; Jacquemin, P.; Garcia, F.; Gonzales, G.; Lopez, J. M.; Ortiz, J. & Solis, A. (1983). On the fishery and biology ofthe giant squid Dosidicus gigas in the Gulf of California, Mexico. *FAO Fisheries Technical Paper*, Vol.231, (February 1983), pp. (306-340), 0429-9345.

El Ghaouth, A.; Arul, J.; Grenier, J. & Asselin, A. (1992). Antifungal activity of chitosan on two post harvest pathogens of strawberry fruits. *Mycological Research*, Vol.96, No.9, (September 1992), pp. (769-779), 0953-7562.

Engel, J. & Bachinger, H. P. (2005). Structure, stability and folding of the collagen triple helix. *Collagen*, Vol.247, (October 2005), pp. (7-33), 0340-1022.

Exposito, J. Y.; Cluzel, C.; Garrone, R. & Lethias, C. (2002). Evolution of collagens. *Anatomical Record*, Vol.268, No.3, (November 2002), pp. (302-316), 0003-276X.

Fernández, M.; Heinämäki, J.; Krogars, K.; Jörgensen, A.; Karjalainen, M.; Colarte, A. & Yliruusi, J. (2004). Solid-State and Mechanical Properties of Aqueous Chitosan-Amylose Starch Films Plasticized With Polyols. *Aaps Pharmscitech*, Vol.5, No.1, (December 2004), pp. (109-114), 1530-9932.

Fornes, F.; Almela, V.; Abad, M. & Agusti, M. (2005). Low concentrations of chitosan coating reduce water spot incidence and delay peel pigmentation of Clementine mandarin fruit. *Journal of the Science of Food and Agriculture*, Vol.85, No.7, (May 2005), pp. (1105-1112), 0022-5142.

Friess, W. (1998). Collagen - biomaterial for drug delivery. *European Journal of Pharmaceutics and Biopharmaceutics*, Vol.45, No.2, (March 1998), pp. (113-136), 0939-6411.

Garcia, J. A.; de Damborenea, J.; Navas, C.; Arenas, M. A. & Conde, A. (2007). Corrosion-erosion of TiN-PVD coatings in collagen and cellulose meat casing. *Surface & Coatings Technology*, Vol.201, No.12, (March 2007), pp. (5751-5757), 0257-8972.

Giménez, B.; Gómez-Estaca, J.; Alemán, A.; Gómez-Guillén, M. C. & Montero, M. P. (2009). Physico-chemical and film forming properties of giant squid (*Dosidicus gigas*) gelatin. *Food Hydrocolloids*, Vol.23, No.3, (July 2009), pp. (585-592), 0268-005X.

Gómez-Estaca, J.; Gómez-Guillén, M. C.; Fernández-Martín, F. & Montero, P. (2011). Effects of gelatin origin, bovine-hide and tuna-skin, on the properties of compound gelatin-chitosan films. *Food Hydrocolloids*, Vol.25, No.6, (January 2011), pp. (1461-1469), 0268-005X.

Gomez-Guillen, M. C.; Ihl, M.; Bifani, V.; Silva, A. & Montero, P. (2007). Edible films made from tuna-fish gelatin with antioxidant extracts of two different murta ecotypes leaves (*Ugni molinae Turcz*). *Food Hydrocolloids*, Vol.21, No.7, (October 2007), pp. (1133-1143), 0268-005X.

Gomez-Guillen, M. C.; Turnay, J.; Fernandez-Diaz, M. D.; Ulmo, N.; Lizarbe, M. A. & Montero, P. (2002). Structural and physical properties of gelatin extracted from different marine species: a comparative study. *Food Hydrocolloids*, Vol.16, No.1, (January 2002), pp. (25-34), 0268-005X.

Horn, M. M.; Amaro Martins, V. C. & de Guzzi Plepis, A. M. (2009). Interaction of anionic collagen with chitosan: Effect on thermal and morphological characteristics. *Carbohydrate Polymers*, Vol.77, No.2, (January 2009), pp. (239-243), 0144-8617.

Hu, J. L.; Meng, Q. H.; Ho, K. C.; Ji, F. L. & Chen, S. J. (2009). The Shape Memory Properties of Biodegradable Chitosan/Poly(L-lactide) Composites. *Journal of Polymers and the Environment*, Vol.17, No.3, (September 2009), pp. (212-224), 1566-2543.

Janjarasskul, T. & Krochta, J. M. (2010). Edible Packaging Materials. *Annual Review of Food Science and Technology, Vol 1*, Vol.1, (January 2010), pp. (415-448), 1941-1413.

Jongjareonraka, A.; Benjakula, S.; Visessanguanb, W.; Nagaic, T. & Tanakad, M. (2005). Isolation and characterisation of acid and pepsin-solubilised collagens from the skin of Brownstripe red snapper (*Lutjanus vitta*). *Food Chemistry*, Vol.93, No.3, (December 2005), pp. (475-484), 0021-8561.

Kim, S. K. & Mendis, E. (2006). Bioactive compounds from marine processing byproducts - A review. *Food Research International*, Vol.39, No.4, (October 2006), pp. (383-393), 0963-9969.

Kim, S. K.; Mendis, E.; Rajapakse, N. & Byun, H. G. (2005). Investigation of jumbo squid (*Dosidicus gigas*) skin gelatin peptides for their in vitro antioxidant effects. *Life Sciences*, Vol.77, No.17, (September 2005), pp. (2166-2178), 0024-3205.

Kolodziejska, I.; Sikorski, Z. E. & Niecikowska, C. (1999). Parameters affecting the isolation of collagen from squid (Illex argentinus) skins. *Food Chemistry*, Vol.66, No.2, (August 1999), pp. (153-157), 0308-8146.

Kurita, K. (2006). Chitin and chitosan: Functional biopolymers from marine crustaceans. *Marine Biotechnology*, Vol.8, No.3, (June 2006), pp. (203-226), 1436-2228.

Le-Tien, C.; Millette, M.; Lacroix, M. & Mateescu, M. A. (2004). Modified alginate and chitosan for lactic acid bacteria immobilization. *Biotechnology Appied Biochemistry*, Vol.39, No.3, (June 2004), pp. (347–354), 0273-2289.

Lee, C. H.; Singla, A. & Lee, Y. (2001). Biomedical applications of collagen. *International Journal of Pharmaceutics*, Vol.221, No.1-2, (June 2001), pp. (1-22), 0378-5173.

Lee, C. R.; Grodzinsky, A. J. & Spector, M. (2001). The effects of cross-linking of collagen-glycosaminoglycan scaffolds on compressive stiffness, chondrocyte-mediated contraction, proliferation and biosynthesis. *Biomaterials*, Vol.22, No.23, (Dec 2001), pp. (3145-3154), 0142-9612.

Lee, H. G.; Lee, K. Y. & Shim, J. (2004). Mechanical properties of gellan and gelatin composite films. *Carbohydrate Polymers*, Vol.56, No.2, (June 2004), pp. (251-254), 0144-8617.

Lee, S. J.; Lee, J. Y.; Kim, K. H.; Shin, S. Y.; Rhyu, I. C.; Lee, Y. M.; Park, Y. J. & Chung, C. P. (2006). Enhanced bone formation by transforming growth factor-beta 1-releasing collagen/chitosan microgranules. *Journal of Biomedical Materials Research Part A*, Vol.76A, No.3, (March 2006), pp. (530-539), 1549-3296.

Li, H. Y. & Yu, T. (2001). Effect of chitosan on incidence of brown rot, quality and physiological attributes of postharvest peach fruit. *Journal of the Science of Food and Agriculture*, Vol.81, No.2, (January 2001), pp. (269-274), 0022-5142.

Li, Q.; Dunn, E. T.; Grandmaison, E. W. & Goosen, M. F. A. (1992). Applications and Properties of Chitosan. *Journal of Bioactive and Compatible Polymers*, Vol.7, No.4, (November 1992), pp. (370-397), 08839115.

Liang, Z. H.; Zhou, C. R.; Zhang, Z. Y.; Li, L. H. & Wang, B. (2005). Microstructure of collagen and chitosan blend films. *Chinese Journal Applied Chemistry*, Vol.22, (February 2005), pp. (62-72), 1000-0518.

Lima, C. G. A.; de Oliveira, R. S.; Figueiró, S. D.; Wehmann, C. F.; Góes, J. C. & Sombra, A. S. B. (2006). DC conductivity and dielectric permittivity of collagen-chitosan films. *Materials Chemistry and Physics*, Vol.99, No.2-3, (December 2006), pp. (284-288), 0254-0584,

Luna-Raya, M. C.; Urciaga-Garcia, J. I.; Salinas-Zavala, C. A.; Cisneros-Mata, M. A. & Beltran-Morales, L. F. (2006). Diagnóstico del consumo del calamar gigante en México y en Sonora. *Economía, Sociedad y Territorio*, Vol.6, No.22, (June 2006), pp. (535-560), 1405- 8421.

Markaida, U.; Rosenthal, J. J. C. & Gilly, W. F. (2005). Tagging studies on the jumbo squid (*Dosidicus gigas*) in the Gulf of California, Mexico. *Fishery Bulletin*, Vol.103, No.1, (March 2005), pp. (219-226)., 0090-0656.

Martínez-Camacho, A. P.; Cortez-Rocha, M. O.; Castillo-Ortega, M. M.; Burgos-Hernández, A.; Ezquerra-Brauer, J. M. & Plascencia-Jatomea, M. (2011). Antimicrobial activity of chitosan nanofibers obtained by electrospinning. *Polymer International*, (2011), DOI 10.1002, 0959-8103.

Meena, C.; Mengi, S. A. & Deshpande, S. G. (1999). Biomedical and industrial applications of collagen. *Proceedings of the Indian Academy of Sciences-Chemical Sciences*, Vol.111, No.2, (April 1999), pp. (319-329), 0253-4134.

Miller, E. J. & Rhodes, R. K. (1982). Preparation and Characterization of the Different Types of Collagen. *Methods in Enzymology*, Vol.82, (November 1982), pp. (33-64), 0076-6879.

Mo, X. M.; Chen, Z. G.; He, C. L. & Wang, H. S. (2008). Intermolecular interactions in electrospun collagen-chitosan complex nanofibers. *Carbohydrate Polymers*, Vol.72, No.3, (May 2008), pp. (410-418), 0144-8617.

Montero, P.; Morales, J. & Moral, A. (2000). Isolation and partial characterization of two types of muscle collagen in some cephalopods. *Journal of Agricultural and Food Chemistry*, Vol.48, No.6, (June 2000), pp. (2142-2148), 0021-8561.

Nagai, T. (2004). Characterization of collagen from Japanese sea bass caudal fin as waste material. *European Food Research and Technology*, Vol.218, No.5, (April 2004), pp. (424-427), 1438-2377.

Nalwa, H., Ed. (2005). *Handbook of Nanostructured Biomaterials and Their Applications in Nanobiotechnology. II.* Stevenson Ranch, California, American Scientific Publishers.

Nesis, K. N. (1985). *Oceanic Cephalopods: Distribution, Life Forms, Evolution* Nauka, Moscow

Nigmatullin, C. M.; Nesis, K. N. & Arkhipkin, A. I. (2001). A review of the biology of the jumbo squid Dosidicus gigas (*Cephalopoda: Ommastrephidae*). *Fisheries Research*, Vol.54, No.1, (December 2001), pp. (9-19), 0165-7836.

No, H. K.; Kim, S. H. & Prinyawiwatkul, W. (2007). Effect of molecular weight, type of chitosan, and chitosan solution pH on the shelf-life and quality of coated eggs. *Journal of Food Science*, Vol.72, No.1, (February 2007), pp. (S44-S48), 0022-1147.

Nogueira, M. G.; Ferreira, C. R.; Cárdenas, G. & Inocentinni, L. H. (2005). Effects of Neutralization Process on Preparation and Characterization of Chitosan Membranes for Wound Dressing. *Macromolecular Symposia*, Vol. 229, No.1, (November 2005), pp. (253-257), 1521-3900.

O'Sullivan, M.; Longares, A.; Monahan, F. J. & O'Riordan, E. D. (2005). Physical properties of edible films made from mixtures of sodium caseinate and WPI. *International Dairy Journal*, Vol.15, No.12, (December 2005), pp. (1255-1260), 0958-6946.

Ohkawa, K.; Cha, D. I.; Kim, H.; Nishida, A. & Yamamoto, H. (2004). Electrospinning of chitosan. *Macromolecular Rapid Communications*, Vol.25, No.18, (September 2004), pp. (1600-1605), 1022-1336.

Oxford University. (1893). *The Little Oxford Dictionary* (8th Edition), Oxford University Press 0198611285, Oxford, United Kingdom.

Pelissari, F. M.; Grossmann, M. V. E.; Yamashita, F. & Pineda, E. A. G. (2009). Antimicrobial, Mechanical, and Barrier Properties of Cassava Starch-Chitosan Films Incorporated with Oregano Essential Oil. *Journal of Agricultural and Food Chemistry*, Vol.57, No.16, (August 2009), pp. (7499-7504), 0021-8561.

Plascencia-Jatomea, M.; Martinez-Camacho, A. P.; Cortez-Rocha, M. O.; Ezquerra-Brauer, J. M.; Graciano-Verdugo, A. Z.; Rodriguez-Felix, F.; Castillo-Ortega, M. M. & Yepiz-Gomez, M. S. (2010). Chitosan composite films: Thermal, structural, mechanical and antifungal properties. *Carbohydrate Polymers*, Vol.82, No.2, (September 2010), pp. (305-315), 0144-8617.

Privalov, P. L. & Tiktopoulo, E. I. (1970). Thermal conformation transformation of tropocollagen. Part I. Calorimetric study. *Biopolymers*, Vol.9, No.2, (February 1970), pp. (127-253), 1097-0282.

Quereshi, S.; Mhaske, A.; Raut, D.; Singh, R.; Mani, A. & Patel, J. (2010). Extraction and partial characterization of collagen from different animal skins. *Biotechnology*, Vol.2, No.9, (January 2010), pp. (28-31), 2076-5061.

Raafat, D. & Sahl, H. G. (2009). Chitosan and its antimicrobial potential - a critical literature survey. *Microbial Biotechnology*, Vol.2, No.2, (March 2009), pp. (186-201), 1751-7907.

Sato, K.; Yoshinaka, R.; Itoh, Y. & Sato, M. (1989). Molecular species of collagen in the intramuscular connective tissue of fish. *Comparative Biochemistry and Physiology*, Vol. 92B, No. 1, (January, 1989), pp. (87-91), 1096-4959

Schauer, C. L. & Schiffman, J. D. (2008). A review: Electrospinning of biopolymer nanofibers and their applications. *Polymer Reviews*, Vol.48, No.2, (January 2008), pp. (317-352), 1558-3724.

Schnepf, M.; Wu, Y.; Rhim, J. W.; Weller, C. L.; Hamouz, F. & Cuppett, S. (2000). Moisture loss and lipid oxidation for precooked beef patties stored in edible coatings and films. *Journal of Food Science*, Vol.65, No.2, (March 2000), pp. (300-304), 0022-1147

Shahidi, F. (2006). *Maximising the value of marine by-products* CRC Press, 9780849391521, St. Johns, Canada

Shahidi, F.; Arachchi, J. K. V. & Jeon, Y. J. (1999). Food applications of chitin and chitosans. *Trends in Food Science & Technology*, Vol.10, No.2, (February 1999), pp. (37-51), 0924-2244

Shahidi, F.; Jeon, Y. J. & Kamil, J. Y. V. A. (2002). Chitosan as an edible invisible film for quality preservation of herring and Atlantic cod. *Journal of Agricultural and Food Chemistry*, Vol.50, No.18, (August 2002), pp. (5167-5178), 0021-8561

Shane, K. & Champa, A. (2009). Rapid biomineralization of chitosan microparticles to apply in bone regeneration. *Journal of Materials Science*, Vol.21, No.2, (July 2009), pp. (393-398), 0022-2461.

Shanmugasundaram, N.; Ravichandran, P.; Neelakanta Reddy, P.; Ramamurty, N.; Pal, S. & Panduranga Rao, K. (2001). Collagen-chitosan polymeric scaffolds for the in vitro culture of human epidermoid carcinoma cells. *Biomaterials*, Vol.22, No.14, (June 2001), pp. (1943-1951), 0142-9612.

Sharma, C. P.; Pillai, C. K. S. & Paul, W. (2009). Chitin and chitosan polymers: Chemistry, solubility and fiber formation. *Progress in Polymer Science*, Vol.34, No.7, (July 2009), pp. (641-678), 0079-6700.

Shen, X. R.; Kurihara, H. & Takahashi, K. (2007). Characterization of molecular species of collagen in scallop mantle. *Food Chemistry*, Vol.102, No.4, (September 2007), pp. (1187-1191), 10.1016

Shi, D.-h.; Cai, D.-z.; Zhou, C.-r.; Rong, L.-m.; Wang, K. & Xu, Y.-c. (2005). Development and potential of a biomimetic chitosan/type II collagen scaffold for cartilage tissue engineering. *Chinese Medical Journal*, Vol.118, No.17, (January 2005), pp. (1436-1443), 0366-6999.

Sikorski, Z. E. & Kolodziejska, I. (1986). The Composition and Properties of Squid Meat. *Food Chemistry*, Vol.20, No.3, (October 1986), pp. (213-224), 0308-8146.

Sikorski, Z.E, Sun Pan, B. & Shahidi, F. (1994). *Seafood proteins*. Chapman & Hall, ISBN 9780412984815 New York, USA.

Sionkowska, A. (2006). The influence of UV light on collagen/poly(ethylene glycol) blends. *Polymer Degradation and Stability*, Vol.91, No.2, 2006), pp. (305-312), 0141-3910

Sionkowska, A. (2011). Current research on the blends of natural and synthetic polymers as new biomaterials: Review. *Progress in Polymer Science*, Vol.36, No.9, (May 2011), pp. (1254-1276), 0079-6700.

Sionkowska, A.; Wisniewski, M.; Skopinska, J.; Poggi, G. F.; Marsano, E.; Maxwell, C. A. & Wess, T. J. (2006). Thermal and mechanical properties of UV irradiated collagen/chitosan thin films. *Polymer Degradation and Stability*, Vol.91, No.12, (December 2006), pp. (3026-3032), 0141-3910.

Suyatma, N. E.; Copinet, A.; Tighzert, L. & Coma, V. (2005). Effects of hydrophilic plasticizers on mechanical, thermal and surface properties of chitosan films. *Journal of Agricultural and Food Chemistry*, Vol.53, No.7, (April 2005), pp. (3950−3957), 0021-8561.

Taipe, A.; Yamashiro, C.; Mariategui, L.; Rojas, P. & Roque, C. (2001). Distribution and concentrations of jumbo flying squid (*Dosidicus gigas*) off the Peruvian coast between 1991 and 1999. *Fisheries Research*, Vol.54, No.1, (December 2001), pp. (21-32), 0165-7836.

Tangsadthakun, C.; Damrongsakkul, S.; Kanokpanont, S.; Sanchavanakit, N.; Pichyangkura, R.; Banaprasert, T. & Tabata, Y. (2007). The influence of molecular weight of chitosan on the physical and biological properties of collagen/chitosan scaffolds. *Journal of Biomaterials Science-Polymer Edition*, Vol.18, No.2, (February 2007), pp. (147-163), 0920-5063.

Tapia-Blacido, D.; Sobral, P. J. & Menegalli, F. C. (2007). Development and characterization of biofilms based on amaranth flour (*Amaraanthus caudatus*). *Journal of Food Engineering*, Vol.67, (March 2007), pp. (215-223), 0260-8774.

Taravel, M. N. & Domard, A. (1993). Relation between the physicochemical characteristics of collagen and its interactions with chitosan: I. *Biomaterials*, Vol.14, No.12, (October 1993), pp. (930-938), 0142-9612.

Taravel, M. N. & Domard, A. (1995). Collagen and its interaction with chitosan: II. Influence of the physicochemical characteristics of collagen. *Biomaterials*, Vol.16, No.11, (December 1995), pp. (865-871), 0142-9612.

Tharanathan, R. N. (2003). Biodegradable films and composite coatings: Past, present, and future. *Trends in Food Science and Technology*, Vol.14, 2003), pp. (71-78), 0924-2244.

Tharanathan, R. N. & Srinivasa, P. C. (2007). Chitin/chitosan - Safe, ecofriendly packaging materials with multiple potential uses. *Food Reviews International*, Vol.23, No.1, (March 2007), pp. (53-72), 8755-9129.

Tharanathan, R. N.; Srinivasa, P. C.; Baskaran, R.; Ramesh, M. N. & Prashanth, K. V. H. (2002). Storage studies of mango packed using biodegradable chitosan film. *European Food Research and Technology*, Vol.215, No.6, (December 2002), pp. (504-508), 1438-2377.

Tohni, E. (2002). Obtention and characterization of collagen-chitosan blends. *Quimica Nova*, Vol.25, No.6, (January 2002), pp. (943-948), 0100-4042

Torres-Arreola, W.; Pacheco-Aguilar, R.; Sotelo-Mundo, R. R., Rouzaud-Sandez, O. & Ezquerra-Brauer, J. M. (2008). Partial Characterization of Collagen from Mantle, Fin, and Arms of Jumbo Squid (*Dosidicus Gigas*). *Ciencia Y Tecnologia Alimentaria*, Vol.6, No.2, (December 2008), pp. (101-108), 1135-8122.

Uriarte-Montoya, M.; Arias-Moscoso, J.; Plascencia-Jatomea, M.; Santacruz-Ortega, H.; Rouzaud-Sández, O.; Cardenas-Lopez, J.; Marquez-Rios, E. & Ezquerra-Brauer, J. (2010). Jumbo squid (*Dosidicus gigas*) mantle collagen: Extraction, characterization, and potential application in the preparation of chitosan–collagen biofilms. *Bioresource Technology*, Vol.10, No.11, (January 2010), pp. (4212-4219), 0960-8524.

Usha, R. & Ramasami, T. (2004). The effects of urea and n-propanol on collagen denaturation: using DSC, circular dicroism and viscosity. *Thermochimica Acta*, Vol.409, No.2, (January 2004), pp. (201-206), 0040-6031.

Venugopal, V. (2009). *Marine Products for Healthcare: Functional and Bioactive Nutraceutical Compounds from the Ocean (Functional Foods and Nutraceuticals)* Taylor & Francis Group, 978-1420052633, New York, USA.

Voet, D. & Voet, J. (1995). *Biochemistry* John Wiley & Sons, Inc., 0-471-58651-x, New York.

Waluda, C. M.; Yamashiro, C.; Elvidge, C. D.; Hobson, V. R. & Rodhouse, P. G. (2004). Quantifying light-fishing for *Dosidicus gigas* in the Eastern Pacific using satellite remote sensing. *Remote Sensing of Environment*, Vol.91, No.2, (May 2004), pp. (129-133), 0034-4257.

Wang, L. & Stegemann, J. P. (2011). Thermogelling chitosan and collagen composite hydrogels initiated with β-glycerophosphate for bone tissue engineering. *Biomaterials*, Vol.31, No.14, (February 2011), pp. (3976-3985), 0142-9612.

Wang, X. H.; Yan, Y. N.; Lin, F.; Xiong, Z.; Wu, R. D.; Zhang, R. J. & Lu, Q. P. (2005). Preparation and characterization of a collagen/chitosan/heparin matrix for an implantable bioartificial liver. *Journal of Biomaterials Science-Polymer Edition*, Vol.16, No.9, (September 2005), pp. (1063-1080), 0920-5063.

Wess, T. J.; Sionkowka, A.; Wisniewski, M.; Skopinska, J. & Kennedy, C. J. (2004). Molecular interactions in collagen and chitosan blends. *Biomaterials*, Vol.25, No.5, (February 2004), pp. (795-801), 0142-9612.

Westbroek, P.; De Vrieze, S.; Van Camp, T. & Van Langenhove, L. (2007). Electrospinning of chitosan nanofibrous structures: feasibility study. *Journal of Materials Science*, Vol.42, No.19, (October 2007), pp. (8029-8034), 0022-2461.

Yu, L.; Dean, K. & Li, L. (2006). Polymer blends and composites from renewable resources. *Progress in Polymer Science*, Vol.31, No.6, (June 2006), pp. (576-602), 0079-6700.

Zhang, M.; Li, X. H.; Gong, Y. D.; Zhao, N. M. & Zhang, X. F. (2002). Properties and biocompatibility of chitosan films modified by blending with PEG. *Biomaterials*, Vol.23, No.13, (July 2002), pp. (2623-2826), 2641-2648.

Zhang, Q.; Liu, L.; Ren, L. & Wang, F. (1997). Preparation and characterization of collagen-chitosan composites. *Journal of Applied Polymer Science*, Vol.64, No.11, (December 1997), pp. (2057–2263), 1097-4628.

Zhang, Y. W.; Wang, Y. S.; Wu, C. X. & Zhao, J. X. (2007). Studies on preparation and properties of PAA/gelatin core-shell nanoparticles via template polymerization. *Polymer*, Vol.48, No.20, (September 2007), pp. (5950-5959), 0032-3861.

3

Pharmaceutically Used Plasticizers

Eva Snejdrova and Milan Dittrich
Faculty of Pharmacy,
Charles University in Prague
Czech Republic

1. Introduction

The extensive use of polymers in medical and pharmaceutical applications including particularly packaging, medical devices, drug carriers and coatings has caused a substantial demand for the proper plasticizers. Although there are many plasticizers used in the chemical industry, only a few of them have been approved for pharmaceutical applications. The natural-based plasticizers characterized by low toxicity and low migration are required nowadays not only for pharmaceutical and medical applications. In this respect, most of traditional plasticizers are not applicable in this area.

External plasticizers added to pharmaceutically used polymers interact with their chains, but are not chemically attached to them by primary bonds therefore their lost by evaporation, migration or extraction is possible. The benefit of using external plasticizers is the chance to select the right plasticizer type and concentration depending on the desired therapeutic system properties particularly drug release. Low volatile substances with average molecular weights between 200 and 400 such as diesters derived from dicarboxylic acids (e.g. sebacic acid, azelaic acid) or from ethylene glycol and propylene glycol, citric acid (tributylcitrate, triethylcitrate) or glycerol (triacetin, tributyrin) are used. Plasticizer as minor component of polymeric drug delivery systems has not been strictly defined. Even liquid drugs or liquids with a potential pharmacodynamic effect can serve as plasticizers.

As well structural water in the hydrophilic polymer seems to be an internal plasticizer of the polymeric drug delivery systems. In case of contact with body fluids after application the hydrophilic plasticizer can be released from polymer and thus conditions for the incorporated drug release are changed. The hydrophobic plasticizer remains in the system and ensures standard conditions during the process of drug release. On the other hand, hydrophilic plasticizer added to the polymeric drug carrier in high concentration can lead to an increase in water diffusion into the polymer, thus diffusivity parameters of the system are changed. As a consequence the kinetics of drug release is changed by elimination of lag time of drug release process. Plasticizers decrease viscosity and thus can enable or facilitate the application of some preparations; e.g. sufficient low viscosity at temperature bellow 50 °C is necessary for easy manipulation and simply and harmless application of implants in situ via an injection needle or trocar device.

2. Pharmaceutically used plasticizers

The primary role of all plasticizers as low molecular weight non-volatile additives is to improve the flexibility and processability of polymers by lowering the second order transition temperature (glass transition temperature, T_g) (Rosen, 1993). The extent of T_g reduction in the presence of a plasticizer can be used as a parameter to assess the plasticization efficiency (Senichev & Tereshatov, 2004).

When incorporated into a polymeric material, a plasticizer improves the workability and flexibility of the polymer by increasing the intermolecular separation of the polymer molecules. This results in a reduction in elastic modulus, tensile strength, polymer melt viscosity and T_g. The polymer toughness and flexibility is improved and lower thermal processing temperatures can be employed. (Zhu et al., 2002).

The attributes of an ideal plasticizer are changed with each application. When selecting an appropriate plasticizer, the compatibility with the polymer and plasticization efficiency are the pivotal criteria. Incompatibility is commonly evidenced by phase separation between the biopolymer and plasticizer, presented in the form of exudated drops on the surface of the product immediately after its blending or during final application (Wilson, 1995).

Historically the first plasticized polymer used as a medicinal preparation since the 19th century has been cellulose nitrate. Its solution in a mixture of ethanol and diethyl ether is called collodion and used to cover wounds. After administration, the solvents quickly evaporate and cause unpleasant tension. That is why its plasticizing with 5% castor oil was introduced (Murray, 1867). The composition after incorporation of an antimicrobial substance functions as a protective barrier.

2.1 Criterions of the plasticizer selection in medicine and pharmacy

Different requirements are important for the choice of a plasticizer for polymeric dosage forms in comparison with these for the technical plasticization. Pharmaceutically used plasticizers are selected according these criteria in the following order of importance (Rahman & Brazel, 2004):

- biocompatibility
- compatibility of a plasticizer with a given polymer
- effect of plasticizer on drug release
- effect of plasticizer on mechanical properties
- processing characteristics
- cost-benefit analysis.

2.1.1 Toxicity

In the selection of a plasticizer suitable for the formulation of dosage forms a great emphasis is laid on the criterion of toxicity. This problem is solved with regard to the mode of administration, dosing frequency and dosage size. For example, the tablet coating contains a plasticizer in the order of milligrams, whereas the implants *in situ* may contain hundreds of milligrams of plasticizers.

In the EU there exists the obligatory document ICH Topic Q3C titled Impurities: Guideline for Residual Solvents, which distinguishes three classes of solvents. Class 1 must not be used, Class 2 has a limited daily dose and concentration, and Class 3 includes solvents accepted in usual amounts in pharmaceutical products. The Class 3 residual solvents are accepted without marked restrictions in daily doses of up to 50 mg or in concentrations up to 0.5 %. Class 3 solvents are listed in Table 1 (European Medicines Agency, 2010). Many of the above mentioned solvents can be used as plasticizers of polymers which should be limited by GMP or pharmacopoeial requirements. The selection of plasticizers appropriate for dosage forms formulation is shown in their list in the 35th edition of the United States Pharmacopoeia (USP 35, 2011) (Table 2).

Acids	Acetic acid, Formic acid
Alcohols	1-Butanol, 2-Butanol, Ethanol, 2-Methyl-1-butanol, 2-Methyl-1-propanol, 1-Pentanol, 1-Propanol, 2-Propanol
Esters	Ethyl acetate, Ethyl formate, Isopropyl acetate, Methyl acetate, Propyl acetate
Ethers	Anisole, tert-Butylmethyl ether, Ethyl ether
Hydrocarbons	Cumene, Heptane, Pentane
Ketones	Acetone, Methylethyl ketone, Methylisobutyl ketone
Others	Dimethyl sulfoxide

Table 1. Class 3 solvents with low toxic potential.

Hydrophilic	Hydrophobic
Glycerin	Acetyl Tributyl Citrate
Polyethylene Glycols	Acetyl Triethyl Citrate
Polyethylene Glycol Monomethyl Ether	Castor Oil
Propylene Glycol	Diacetylated Monoglycerides
Sorbitol Sorbitan Solution	Dibutyl Sebacate
	Diethyl Phthalate
	Triacetin
	Tributyl Citrate
	Triethyl Citrate

Table 2. List of plasticizers declared in USP 35-NF 30.

As there is no single universal mechanism of polymer plasticization, there is no universal criterion for its selection and for the evaluation of its efficacy. The use of internal plasticizers based on modifications of monomer units by decreasing the polarity of the groups, or modification with large side groups is problematic in pharmacy. The main cause is the fact that an introduction of new compounds into pharmacopoeias as the principal pharmaceutical standards is as complicated as an introduction of a new active substance. It is a process that lasts several years and is considerably costly.

On the rule, external plasticizers are used, which are less heterogeneously miscible, more frequently molecularly miscible with an amorphous phase of polymers. They are the solvents or thinners used in a minority concentration in a mixture with a polymer. The range of the concentrations of the plasticizer in the polymer is due to, besides other things, the crystallinity of the polymer, usually 5 % to 30 %, but there exist also deviations from this range.

When plasticizers are present in low concentrations, their effect is often the exact opposite of what is typically expected. Low concentrations of plasticizer often result in an increase in the rigidity of the polymer instead of the expected softening effect. This effect is known as antiplasticization, or effect of low plasticizer concentrations (Chamarthy & Pinal, 2008).

It is of advantage when plasticizers are low volatile. The size of the molecule of plasticizers includes a wide range from a water molecule to the molecules of low-molecular oligomers and polymers. Schematically, hydrophilic and hydrophobic plasticizers are distinguished; the border between them is not officially declared, the classification is subjective.

Because of the prevalence of fully biodegradable polymers, particularly for parenteral controlled-release systems as they do not require surgical retrieval from the body after completion of the drug release, recent research has focused on developing compatible plasticizers that also biodegrade.

2.1.2 Biocompatibility of pharmaceuticaly used plasticizers

The use of any material in the formulation of biomedical and medicinal devices and preparations must comply with three requirements (Zeus Industrial Products, 2005):

- It must be biocompatible from the standpoint of specific use.
- It must comply with the complex legislative and regulatory rules.
- It must comply with environmental requirements.

Biocompatibility is a general term which describes the suitability of the material in the exposure to body fluids. The material in a specific application is considered to be biocompatible if it makes it possible for the organism to function without complications, such as allergic reactions or other side effects. The complications may include chronic inflammations in contact or due to the effect of eluted components interacting with the organism, then the formation of substances which are cytotoxic, substances inducing skin irritation, restenosis, thrombosis, etc. Biocompatibility concerns medical devices and pharmaceutical products.

Various tests are used in dependence on the type of the medical device and the mode of its administration. Medical devices sold in the EU must comply with the EU Medical Device Directive 93/42/EEC, in the implemented form as 2007/47/EC, which harmonizes the legislation in the EU. At present, ISO 10993 comprises a series of 20 parts for various aspects of biocompatibility testing prior to the clinical study. ISO 14155:2011 concerns good clinical practice for the performance of safe clinical studies on human subjects. Three categories of medical device products are differentiated according to the contact with the human tissue. Limited contact is shorter than 24 hours, the prolonged one is between 24 hours and 30 days, and the permanent contact lasts for more than 30 days. Many plasticized medicated implanted systems are in the concordance with category of medical devices for permanent contact with very demanding testing.

Pharmaceutical products containing a pharmaceutical drug have as their standard for the biocompatibility check international or national pharmacopoeias, American Pharmacopoeia in the study of biocompatibility has been superseded from a major part by the standard ISO 10993. Tests following the USP are carried out on animals in three versions:

- Systemic injection test (intravenous and intraperitoneal)
- Intracutaneous test
- Implantation test

The tests called in the USP as Biological Reactivity Tests are carried out with the pharmaceutical dosage form or the medical device or the extracts obtained from them at different temperatures for different periods of action of the extraction reagent.

2.1.3 Plasticizer uptake by polymer

The rate of plasticizer uptake is depended on the type and concentration of the plasticizer and the type of polymer dispersion, and also on the water solubility of the plasticizer. The plasticization time had a minimal effect on the rate of uptake of water-soluble plasticizers, while it had a strong effect on the uptake of water-insoluble plasticizers. Depending on plasticizer water-solubility and the added amount, when a plasticizer is added to an aqueous polymer dispersion, it is first dissolved and/or dispersed within the outer water phase. Subsequently, the plasticizer diffuses into the polymer particles. A sufficient amount of time for plasticizer uptake by the polymer particles is necessary to avoid forming an inhomogeneous plasticized system (Bodmeier & Paeratakul, 1997).

2.1.4 Plasticizer leaching out of polymer and methods of reduction

Plasticizer migration refers to any method by which a plasticizer leaves a polymer to a solid, liquid or gas phase, which includes solid–solid migration, evaporation of plasticizer, and liquid leaching. These mentioned processes signify the loss of plasticizers from the plasticized polymeric system. The most important way of plasticizer migration within the polymer drug delivery system represents its leaching by physiological fluids after application of the dosage form. Leaching is the major trouble encountered during the plasticizing of polymeric drug delivery systems, as it can eventually result in drastic alteration in all the functions of the plasticizer, and thus the properties of the initially plasticized polymer system, notably the incorporated drug release patterns.

Plasticized polymers used in drug delivery systems come to contact with liquid after application into the body. Plasticizers tend to diffuse down the concentration gradient to the interface between the polymer surface and the external medium. The interfacial mass transport to the surrounding medium has been found to be the limiting step rather than diffusion of the plasticizer through the matrix to the surface. This rate is usually a function of temperature and initial plasticizer concentration (Foldes, 1998). If plasticizers leach out to a liquid, polymers fail to retain their flexibility while the loss of plasticizers leaves the polymers inappropriate for the desired application. Leaching issue is one of the toughest challenges regarding the research of plasticizers today.

The most effective approaches how to reduce the leaching of plasticizers into physiological fluids are particularly surface modifications. Among a variety of surface modification techniques, (i) surface cross-linking, (ii) modification of surface hydrophilicity/lipophilicity by grafting of water soluble polymers to the surface of biomaterials, (iii) surface coating and (iv) surface extraction have been used.

Surface modification of polymers not only prevents the plasticizer from leaching but also improve the biocompatibility of a polymer without compromising. Coating the polymer surface with some non-migrating material may often cause a reduction in flexibility of the polymeric materials due to the thickness of the coating layer. During surface extraction, a material is shortly exposed to some solvent for the plasticizer, and then dried. Subsequently the plasticizer distribution in the polymer is not homogenous and interfacial mass transfer of the plasticizer is blocked with the rigid surface. It follows that the mechanical properties of the polymer system are very often influenced in a negative way.

Leaching resistance of flexible PVC has also been improved by grafting polyethylene glycol, which is often used in biomaterials to prevent biological recognition and protein adhesion. The decrease in plasticizer leaching after polyethylene glycol-grafting is presumably due to the hydrophilic polyethylene glycol surface acting as a barrier to the diffusion of di(2-ethylhexyl) phthalate (DEHP) from the PVC matrix. (Lakshmi & Jayakrishnan 1998). There are radical possibilities of how to solve the plasticizer migration out, namely the use of polymeric or oligomeric plasticizers instead of the low molecular weight ones, or alternative (so called non-classical, non-traditional, multifunction) plasticizers, even alternative polymers which do not require plasticizers.

2.2 The effect of plasticizers on human health

Phthalic acid esters found applications as plasticizers for the first time in 1920s and continue to be the largest class of plasticizers in the 21st century (Rahman & Brazel 2004). DEHP, also known as dioctyl phthalate (DOP), was introduced in 1930s and has been the most widely used plasticizer up to the present time. Thus, the use of plasticizers is being questioned due to their possible toxicity problems, related to their migration out of the polymer. Nowadays, there is increasing interest in the use of natural-based plasticizers that are characterized by low toxicity and low migration. This group includes epoxidized triglyceride vegetable oils from soybean oil, linseed oil, castor-oil, sunflower oil, and fatty acid esters (Baltacioğlu & Balkose 1999).

Currently, there is a trend towards replacing DEHP by either diisononyl phthalate or diisodecyl phthalate, which are higher molecular weight phthalates and therefore are more permanent, have lower solubility and present slower migration rates. Although a total replacement of synthetic plasticizers by natural-based plasticizers is just impossible, at least for some specific applications such a replacement seems obvious and useful.

A number of medical devices as bags, catheters and gloves, intravenous (i.v.) fluid containers and blood bags, medical tubings, are made from the PVC plasticized with the use of DEHP. PVC i.v. bags typically contain 30-40% DEHP by weight; other devices may contain much as 80% DEHP by weight. Because DEHP is not chemically bound to the polymer in a PVC medical device, it can be released into the solutions and blood products transported by these devices.

DEHP leaching from medical plastics was first observed in the late 1960s (Jaeger & Rubin, 1970). Extensive research began after the International Agency for Research on Cancer classified it as 'possibly carcinogenic to humans' in 1980s (Murphy, 2001). The mechanisms by which DEHP may cause various adverse effects in humans are likely to be multiple and variable, and are not well understood. DEHP belongs to a class of chemicals called "peroxisome proliferators". The greater exposure source is particularly relevant for individuals who are ill, and therefore potentially less able to withstand any toxicant exposure (Tickner et al., 1999).

Particular concern has been raised in neonatal care applications because newborns receive among the highest doses of DEHP from blood transfusions, extracorporeal membrane oxygenation and respiratory therapy. (Sjoberg et. al., 1985; Loff et al., 2000).

Infants and children receiving intravenous total parenteral nutrition infused using the typical PVC-DEHP tubing and ethylene vinyl acetate bags with PVC-DEHP connections potentially receive considerable amounts of DEHP every day. DEHP is extracted from the bags and tubing due to the high solubility of DEHP in lipids and DEHP extraction by total parenteral nutrition depends on the lipid content of each total parenteral nutrition preparation and the flow rate (Kambia et al., 2003).

Adults can also be subjected to DEHP exposure from medical plastics. Kambia et al. (Kambia et al., 2001) studied the leachability of DEHP from PVC haemodialysis tubing during maintenance haemodialysis of 10 patients with chronic renal failure. The patient blood obtained from the inlet and the outlet of the dialyzer was analyzed during a 4 h dialysis session. An average DEHP quantity of 123 mg was extracted from tubing during a single dialysis session, of which approximately 27 mg was retained in the patient's body. The detrimental dose for humans has been estimated at 69 mg/kg per day whereas the average daily exposure to DEHP is much lower (2.3–2.8 mg/kg in Europe and 4 mg/kg in the US). (Murphy, 2001). Application of DEHP as plasticizer was found to have adverse effects on the biocompatibility of the plastic materials used in medical devices. Upon contact with blood, albumin is instantly absorbed on the polymer surface, followed by globulin (Baier, 1972).

There are a number of techniques which could help minimize health and environmental problems owing to the use of leachable plasticizers. One simple way to do this is to use alternative flexible polymers (e.g. polyolefins), which require less or no plasticizers, to accomplish some surface modification, or use plasticizers that have less volatility and leachability, and thus low toxicity. Because it is cheap, clear, and flexible, PVC remains the most widely used material by manufacturers and end users of medical bags and tubing.

2.3 Classification of pharmaceuticaly used plasticizers

Pharmaceuticaly used plasticizers are often distinguished into hydrophilic and hydrophobic, or low molecular weight, oligomeric and polymeric. Water insoluble plasticizers have to be emulsified in the aqueous phase of the polymer dispersions. Plasticizers are incorporated in the amorphous phase of polymers while the structure and size of any crystalline part remains unaffected (Fedorko et al., 2003). The water has an exceptional position as the inherent plasticizer of biopolymers, mainly polysaccharides and proteins.

2.3.1 Water as a plasticizer

Plasticization is a concept which can be understood either as a physical phenomenon, or as a technological process (Kozlov & Papkov, 1982). Water is a natural plasticizer of biopolymers as well as their semisynthetic derivatives. A portion of this water is structural water, which has anomal properties (Coyle et al., 1996). Cotton cellulose at 60 % to 70 % crystallinity contains 6 % to 8 % of water, viscose does even more, and gelatin as collagen hydrolyzate contains 5 % to 15 % of water, according to atmospheric humidity.

The water content influences also the properties of synthetic polymers. The polyacrylate polymer Eudragit RS used to coat the pellets with theophylline changes its mechanical and dissolution properties with the relative humidity on storage (Wu & McGinity, 2000). T_g values are decreased in the biodegradable poly(lactide-co-glycolide) in the environment of water vapours up to by 15 °C. Water content was within a range from 0.3 % to 2.6 %. It has been demonstrated that water responsible for plasticization effect was non-freezable and only a small fraction of this water absorbed from the environment caused degradation of the polymer in the same manner as bulk water. In dependence on temperature and concentration, water can act as a plasticizer and an antiplasticizer (Blasi et al., 2005).

2.3.2 Hydrophilic plasticizers

Hydrophilic plasticizers include the compounds which are without limitations or in a sufficient degree miscible with water. They are on the rule substances with very good biocompatibility, some are components of metabolic processes, and others can be easily eliminated from the organism. The most widely used are polyhydric alcohols, in the first place glycerol. The polymers plasticized with hygroscopic compounds receive water from the atmosphere in an increased degree and this water also possesses a plasticizing effect.

The thermoplastic starch is a material interesting for the use in pharmacy (Willet et al., 1997; Liu et al., 2009). It is produced by heating under pressure and under shear from a mixture of native granules and 20 % to 50 % glycerol. This composition named as opened starch was patented for implantation (Van De Wijdeven, 2010). In the temperatures of 150 to 180 °C the granules melt and a plastic amorphous material is produced (Carvalho et al. 2003). Glycerol and xylitol are important plasticizers of starch; in their presence the starch film is flexible regardless of the water content in it (Bader & Göritz, 1994). With the use of 11 % of water and glycerol or xylitol in low concentrations, an antiplasticizing action of polyalcohols on starch was observed; when a concentration of 15 % for glycerol and 20 % for xylitol was achieved, there occurred a significant decrease in T_g. Between the individual plasticizers there occurs competitive plasticization with three types of interactions: starch/plasticizer, plasticizer/water and starch/water (Chaudhary et al., 2011).

The tests for starch plasticization (cassava starch) included glycerol, sorbitol and a mixture of these two polyhydric alcohols in a ratio of 1:1 (Mali et al., 2005) in a concentration of 0, 20, and 40 g/100 g of starch. The combined effect of relative humidity and the plasticizer on the mechanical properties of films was tested. The hydrophilicity of the films of the plasticizer was the decisive factor influencing the affinity to water. The films plasticized with glycerol adsorbed more humidity and more rapidly, and they were more influenced by plasticization from the standpoint of mechanical properties. Glycerol in comparison with sorbitol and a mixture of both was the most effective plasticizer, much reduced the internal hydrogen bonds between the polymer chains and enlarged the internal space in the molecular structure of starch.

Amylose and starch were plasticized with glycerol or xylitol in various concentrations up to 20 %. On the basis of water sorption, the competition of the plasticizer and water under different activities of water was evaluated. Starch interacts with plasticizers and water by changing its crystallinity. The samples of lower concentrations of the plasticizers contain more humidity in the values of the activity of water in a range of 0.11 to 0.65. With a low activity of water there occurs association of amylose and exclusion of the molecules of the plasticizer. With increasing activity of water over 0.55, lower concentrations of the plasticizer exert no effect on a balanced content of water. It was explained by a strong bondings of glycerol and xylitol on starch chains with the development of cross-linking by means of hydroxyl bonds. Water is thus excluded from the polymer matrix. With a lower activity of water, starch binds both the plasticizer and water. Its antiplasticizing limit for glycerol was found between 10 % and 15 %, for xylitol this effect at its concentrations up to 20 % was not demonstrated (Liu et al., 2011).

Chitosan, partially deacetylated chitin is a biopolymer which has been studied very intensively as a potential carrier of active ingredients for more than two decades. Chitosan salts, in particularly chloride, lactate or gluconate are surface active and filmogenic. Elasticity of films can be improved by their plasticizing. Glycerol was used as a plasticizer of chitosan on a 25% concentration. The material was tested as suitable for the preparation of matrices by mechanical kneading as an alternative procedure to the traditional procedures based on the methods of solvent casting (Epure et al., 2011). Sorption of water, changes in crystallinity and thermomechanical properties were examined when the samples with and without the plasticizer were stored in the surroundings with a different relative humidity. Glycerol has been demonstrated to increase the hydrophilic character of films and acts in a plasticizing manner on the mechanical properties. The most suitable for film storage is the environment of the atmosphere with medium values of relative humidity (57 % RH).

The invention provided an orally dissolving capsule comprising pullulan, a plasticizer, and a dissolution enhancing agent (Rajewski & Haslam, 2008). Polyhydric alcohols such as glycerol, propylene glycol, polyvinyl alcohol, sorbitol and maltitol were proposed.

The biodegradable film was prepared from a blend of native rice starch-chitosan with an addition of various plasticizers in concentrations from 20 % to 60 %, using sorbitol, glycerol and polyethylene glycol 400. With an increased activity of water, a higher content of absorbed water was demonstrated. With increasing relative humidity the time of achievement of a balanced concentration of water was prolonged from 13 days to 24 days. Polymer films plasticized with polyethylene glycol 400 did not increase the content of water with increasing atmospheric humidity (Bourtoom, 2008).

Gelatin is a biopolymer produced by hydrolysis of collagen. Its surface activity and ability to form elastic and firm films are used. Its mechanical properties can be improved by adding plasticizers. Permeability of water vapours, mechanical and thermal properties of gelatin films of the gelatin produced from bovine and porcine hides were measured. The films contained 15 g to 65 g of sorbitol per 100 g of gelatin. Permeability of the conditioned films increased with sorbitol content. The origin of gelatin was important from the standpoint of measured parameters only at a sorbitol concentration higher than 25 g/100 g of gelatin. The samples containing 15 g to 35 g of sorbitol/100 g of gelatin heated in the first cycle possessed a marked glass transition followed by a sol-gel transition. With increasing sorbitol concentration, the glass transition was wider, typical of the system with phase separation. To predict T_g values in the function of sorbitol concentration, the model according to Couchman and Karasz for the ternary system was employed (Sobral et al., 2001).

The effect of glycerol in a concentration of 3 – 7 % and sorbitol 4 – 8 % on the permeability of water vapours, humidity content, solubility and optical transparency of films prepared from the protein isolated from pea seeds was investigated. With an increasing glycerol content the permeability of vapours and humidity content in the films was increased, their solubility was not influenced, films plasticized with sorbitol, on the other hand, possessed lower permeability and humidity content and higher solubility. Different behaviour of plasticized films was explained by different hygroscopic plasticizers. A change in the pH value of solutions in the preparation of films from 7.0 to 11.0 did not influence most parameters (Kovalczyk & Baraniak, 2011).

Glycerol serves a number of functions in soft gelatin capsules – it is a humectant, plasticizer, in a higher concentration it serves as a preservative. It influences the helix formation from linear protein chains of gelatin in dependence on concentration. Its effect on the formation of helices decreases to 10 %, and then increases, on storage the degree of organization of the structure grows more in hard capsules than in the soft ones. With an increasing glycerol concentration, the extent of changes in the structure on storage is increased (Hüttenrauch & Fricke, 1984).

2.3.3 Plasticizers with limited miscibility with water

The border between hydrophilic and hydrophobic plasticizers is not sharp, being connected with its solubility in water. Plasticizers which possess solubility in water lower than 10 % are frequently employed for the formulation of dosage forms either in the form of solutions or they are emulsified in the aqueous phase. On the rule they are highly biocompatible esters of dicarboxylic and tricarboxylic acids or glycerol esters. These items are mentioned below. In the selection of a suitable plasticizer of this category of less polar compounds, two principal criteria are taken into consideration, (1) depression of the glass transition temperature and (2) the parameter of solubility. In the next order of importance are the mechanical properties of plasticized polymers, such as decreased strength, decreased elastic modulus and increased elongation at break. Another parameter for the selection for formulation studies is a decrease in the internal stress or the effect on the permeability of the material and for the release of the active ingredient.

Diesters and triesters of acids:

- Triethyl citrate (TEC)
- Tributyl citrate (TBC)
- Acetyl triethyl citrate (ATEC)
- Dibutyl sebacate (DBS)

- Diethyl phthalate (DEP)
- Dibutyl phthalate (DBP)

Diesters and triesters of alcohols:

- Triacetin (TA)
- Vegetable oils
- Fractionated coconut oil
- Acetylated monoglycerides

They are the plasticizers which are added to synthetic polymers with lower polarity in different fields of human activity. Many of them are encountered in foodstuffs. The team of analytical chemists of the Japanese National Institute of Health Sciences carried out an analysis of 93 samples of foodstuffs from the standpoint of the presence of 10 plasticizers (4 phthalates, 3 adipates, 1 sebacate, 1 citrate, and 1 triglyceride) and used as additives to the covers and vessels of various Japanese manufacturers. The method of gas chromatography/mass spectrometry revealed higher concentrations of diacetylauroyl glycerol, which did not originate from contamination with plastics, but it was used in children's food as an additive. Acetyl tributyl citrate was found in the bottles with sake, migrating from the bottle caps seals. This one as well as the other plasticizers were deep below the maximal tolerated concentrations (Tsumura et al., 2002).

Triesters of citric acid are considered to be very safe. They possess very advantageous parameters of biocompatibility. Their acetylated forms are markedly hydrophobic, mainly acetyl tributyl citrate. In acute, short-term, subchronic and chronic testing they are relatively non-toxic. After ocular and dermal administration to rabbits they were non-irritating, in guinea-pigs acetyl triethyl citrate acted as a sensitizer, whereas acetyl tributyl citrate did not. According to Cosmetic Ingredient Review Expert Panel, esters of citric acid are not considered to be sensitizers (Johnson, 2002). After intravenous administration they decrease blood pressure and spasms of intestinal muscles. The compounds were not genotoxic in the tests on bacteria and on mammals, they did not induce tumours.

Dibutyl sebacate is a widely used plasticizer in pharmacy. Its solubility in water is 40 mg/l at 20 °C, it is odourless and colourless. It possesses very favourable thermal characteristics, above -10 °C it is liquid, at 344 °C is boiling point. The toxicological data indicate that this compound is practically non-toxic after oral administration and also non-irritating in dermal contact (Clayton & Clayton, 1993-1994).

Phthalates are effective plasticizers of many polymers. With regard to the fact that in the case of some pharmaceutical applications, in particular in film coating of tablets, they are used in very small amounts they are still in use; it is, above all, dibutyl phthalate (Lowell Center for Sustainable Production, 2011) and diethyl phthalate (World Health Organization, 2003).

Triacetin has been very often used as a plasticizer and a solvent in pharmaceutical and cosmetic products. It has been affirmed as a GRAS product by FDA for human use in the food industry, and it is safe for cosmetic products (Zondlo & Fiume, 2003). After acute short-term oral administration and dermal exposure it is not toxic or mutagenic; it feebly irritates the guinea pig skin and rabbit eye.

2.3.4 Oligomeric and polymeric plasticizers

An advantage of the plasticizers of this type is a decrease in or a full prevention of their migration from materials (Rasal et al., 2010).

Polyesters derived from aliphatic hydroxy acids are compounds which have been very intensively studied and employed as biodegradable and renewable thermoplastic materials with a potential of replacing the conventional polymers based on mineral oil products. These polyesters are used as carriers of active ingredients with a period of release of these substances for weeks to months. They are the products of polymerization of cyclic dimers, lactones via ring opening method, or the substances developed by a polycondensation reaction, e.g. poly(lactic acid), poly(lactide-co-glycolide). They are mostly polymers which in the glassy state have a small elongation at break. For their plasticization highly biocompatible, if possible completely biodegradable compounds are suitable. As the very suitable ones were demonstrated oligoesters or low-molecular polyesters of identical or similar aliphatic hydroxy acids as plasticized polymers (Martin & Avérous, 2001), and polyesteramides were also proposed (Ljungberg et al., 2005). Polyethylene glycols (PEG) are also suitable for these purposes, their miscibility decreases with molecular mass (Baiardo et al., 2003). PEG with a value of M_n 20 000 very effectively plasticized in a 40 % concentration of poly(L-lactic acid) (Kim et al., 2001). PEG in a concentration above 50 % possesses increased crystallinity, an increased module and decreased ductility (Sheth et al., 1997). Polypropylene glycol also exerts a plasticizing effect on poly(L-lactic acid), its effect on a decrease in crystallinity is lower than in PEG (Kulinski et al., 2006). A blend of two plasticizers called multiple plasticizer, triacetin and oligomeric poly(1,3-butanediol), significantly influences the elastic properties and tensile strength (Ren et al., 2006).

2.3.5 Non-traditional plasticizers

It is advantageous to utilize plasticization effect of some pharmaceutical active agents or of some excipients possessing other functions in the formulated composition. In the literature, these are named as non-traditional, non-conventional, or multifunctional plasticizers. Several studies report the plasticization of polymers by ibuprofen, theophylline, salts of metoprolol and chlorpheniramin and other active ingredients. From the auxiliary compounds it is potentially promising the use of many surfactants, preservatives, solvents, cosolvents, desolvating and coacervating agents as plasticizers. These components of pharmaceutical preparations can act by various mechanisms, as lowering of intermolecular and intramolecular interactions, increasing of macromolecular or segmental mobility with the consequence of ameliorated thermal and mechanical properties, distensibility, adhesion, viscosity etc.

Ibuprofen was found to be very effective in plasticizing of the acrylic film. Ibuprofen interacts with the Eudragit RS 30 D polymer through hydrogen bonding. The glass transition temperature of the Eudragit RS 30 D polymer decreased with the increasing levels of ibuprofen in the polymeric film (Wu & McGinity, 2001). Metoprolol tartrate, chlorpheniramine maleate and ibuprofen are efficient plasticizers for Eudragit RS as shown by the thermal and mechanical properties of drug-loaded polymeric film (Siepmann et al., 2006).

The influence of methylparaben, ibuprofen, chlorpheniramine maleate and theophylline on the mechanical properties of polymeric films of Eudragit® RS 30 D was studied. The results demonstrated that the glass transition temperature of the Eudragit® RS 30 D decreased with increasing levels of methylparaben, ibuprofen and chlorpheniramine maleate in the polymeric coatings. The addition of methylparaben to Eudragit® RS 30 D resulted in significant changes in the mechanical properties, making the polymer softer and more flexible. The T_g of the polymer was significantly reduced (Wu & McGinity, 1999).

3. Drug release influenced by plasticizers

Drug release from polymer drug delivery system is modified by the method of their formation, or by using an appropriate polymer or additive, which could also be a plasticizer. Modified release includes delayed release, extended release (prolonged, sustained), and pulsatile release (Chamarthy & Pinal, 2008).

Dosage forms based on polymeric carriers can be classified according to the mechanism of drug release into the following categories: (i) Diffusion-controlled drug release either from a non-porous polymer drug delivery system or (ii) from a porous polymer drug delivery system, and (iii) disintegration controlled systems (Khandare & Haag, 2010). Diffusion of a drug within a non-porous polymer drug delivery system occurs predominantly through the void spaces between polymer chains, and in the case of a porous polymer drug delivery system by diffusion of a drug through a porous or swelling polymer drug delivery system. The plain fact is that the plasticizer type and concentration must influence the drug release as plasticizers reduce polymer-polymer chain secondary bonding, and provide more mobility for the drug. Plasticizer leaching out of the polymer results in pore formation for burst release of the drug. Subsequent release stage of drug is based on diffusion through the dense polymer phase.

Non-biodegradable polymers are characterized by their durability, tissue compatibility, and mechanical strength, which endure under in vivo conditions without erosion or considerable degradation. Polyurethane, poly(ethylene vinyl acetate), and polydimethylsiloxane are examples of polymer films that follow predictable Fickian diffusion or can be modified for linear or near zero order release. One drawback of these non-biodegradable polymer devices is an occasional need for a second surgical procedure to remove the device, which leads to an increased cost and associated discomfort / inconvenience for the patient.

Drug release from biodegradable polymers is depended on the way of erosion and degradation. The most commonly employed class of biodegradable polymers are the polyesters, which consist mainly of poly(caprolactone), poly(lactic acid), poly(glycolic acid) and copolymers of lactic and glycolic acids. Their degradation mechanism is non-enzymatic random hydrolytic chain scission established. Polymer drug delivery system can be classified in bulk-eroding systems, surface-eroding systems, or systems undergoing both surface and bulk erosion. Polyanhydrides and polyorthoesters degrade only at the surface of the polymer, resulting in a release rate that is proportional to the surface area of the drug delivery system. Poly(lactic acid) and poly(lactic-co-glycolic acid) are reported to undergo both surface and bulk erosion which probably disturbs an even rate of drug delivery.

The release profile of the drug from degradable matrices is typically triphasic. The three phases can be summarized by an initial rapid drug release (burst effect) from the matrix surface, followed by a phase where the incorporated drug diffuses more slowly out of the inner bulk matrix and then, the remaining drug release phase due to bulk degradation of the polymer. The extent of the initial burst release can be controlled by plasticizers. High burst release can be minimized by hydrophobic plasticizers; the opposite effect is achieved by hydrophilic plasticizers, which leach out of polymer in the hydrophilic medium.

Biodegradation rate and thus the release of incorporated drugs depend on the polymer molecular weight parameters. Low molecular weight polymers or oligomers are preferred as drug carriers, as their hydrolysis may proceed simultaneously or just somewhat slower than drug release. A low molecular weight poly(D,L-lactic) acid with a linear molecule constitution is not suitable as a carrier due to a lag-time. Star-like copolymers of hydroxy acids with polyhydric alcohols, such as pentaerythritol, mannitol, glucose, polyvinyl alcohol and others are particularly advantageous. These branched carriers unlike the linear are endowed with low gyration radius (Kissel et al. 1991). Drug release from this type of carriers can be modified not only by molecular weight, but also more significantly by the degree of their branching (Pistel et al., 2001).

3.1 Effect of solubility parameters of the plasticizer on drug release

The physicochemical properties, particularly the solubility parameters of the plasticizer and extent of the plasticizer leaching act the major role in the drug release from a plasticized polymer system. The differences in the drug release patterns are observed in the case of using either lipophilic or hydrophilic plasticizers. The lipophilic plasticizers, (e. g. dibutyl sebacate) are shown to remain within the polymeric system upon exposure to the release media, assuring integral and mechanically resistant coatings during drug release. In contrast, hydrophilic plasticizers leached out of the system, resulting either in decreased mechanical resistance and thus cracking, or in facilitated pore formation. As drug release was controlled by diffusion through the intact membrane and/or water-filled cracks (with significantly different diffusion coefficients), the mechanical stability of the polymeric system and the onset of crack formation are of major importance for the resulting drug release profiles.

3.2 Effect of affinity of the plasticizer to the polymer on drug release

Furthermore, the affinity of the plasticizer to the polymer is found to be decisive. The plasticizer redistribution within the polymeric systems during coating, curing and/or storage affects the drug release rate. For instance, dibutyl sebacate has a higher affinity to ethycellulose than to Eudragit RL, resulting in potential redistributions of this plasticizer within the polymeric systems and changes in the release profiles. Importantly, adequate preparation techniques for the coating dispersions and appropriate curing conditions could avoid these effects, providing stable formulations (Bodmeier & Paeratakul, 1997).

3.3 Effect of plasticizer concentration on drug release

The plasticizer concentration has a significant impact on the drug release of a diffusion-controlled drug delivery system. Low concentrations of the plasticizer often result in an increase in the rigidity of the polymer instead of the expected softening effect. This effect, known as antiplasticization, can be used as a formulation strategy which can modulate drug permeability of polymers used in pharmaceutical systems.

The antiplasticizing effect of water on the transport properties of disintegration controlled systems such as tablets is highly relevant during the manufacturing, handling and storage of the product; water does not antiplasticize during drug release. Once in the body, pharmaceutical formulations are subjected to a water saturated environment. Consequently, water will act exclusively as a plasticizer under such conditions (Chamarthy & Pinal, 2007).

The theophylline release profile from soluble starch plasticized with sorbitol exhibits two valleys, which can be explained as a simultaneous plasticizing effect of water penetrating from the dissolution medium and the antiplasticizing effect of sorbitol contained in the formulation (Chamarthy & Pinal, 2008).

Antiplasticization can be expected to significantly affect drug release and thus a factor that has to be taken into consideration in formulation development.

3.4 Effect of drug-polymer or drug-plasticizer interaction on drug release

The drug-polymer or drug-plasticizer interaction within the polymer drug delivery system can significantly influence the drug release profile. For instance, when triacetin was added to indomethacin loaded poly(methyl methacrylate) (PMMA) microspheres, a desired drug release profile lasting 24 h was achieved. Originally biphasic release profile, an initial burst effect from the surface of the microspheres followed by a slower drug release phase was surmounted by addition of a plasticizer. There might be a hydrogen bonds formation between the indomethacin hydroxyl group and PMMA, no interaction between triacetin and indomethacin or PMMA as the effects of secondary bonds was observed. The release enhancement of indomethacin from microspheres was attributed to the physical plasticization effect of triacetin on PMMA and, to some extent, the amorphous state of the drug.

The plasticization effect of triacetin on PMMA increased the diffusivity of indomethacin from PMMA. However, this effect was not dependent on the formation of secondary bonds between triacetin and PMMA. This indicates that the triacetin molecules physically separate the PMMA chains by locating within them (Yuksel et al., 2011).

An example of how drug-polymer interaction can affect the drug release can be piroxicam-loaded Eudragit E film. The drug-polymer interaction occurring between piroxicam and Eudragit E seems to cause a drag effect, leading to a delay of the piroxicam release from the Eudragit E film (Lin et al., 1995).

Similarly, ibuprofen interacts with the Eudragit RS 30 D polymer through hydrogen bonding, thus ibuprofen acts both as the active ingredient and as the plasticizer for the polymer also. The glass transition temperature of the Eudragit RS 30 D polymer decreased with increasing levels of ibuprofen in the polymeric film. The drug release rate was reduced by increasing the amount of ibuprofen in the polymeric film and by increasing the coating level on the coated beads (Wu & McGinity, 2001).

3.5 Effect of plasticization technology on drug release

Drug release profile can be modified by the preplasticization step, which is often necessary when incorporating plasticizer into the formulation in order to achieve uniform mixing of the polymer and plasticizer, to effectively reduce the polymer T_g, and to lower the processing temperatures. For instance, citric acid monohydrate combined with triethyl citrate in the powder blend was found to plasticize Eudragit S 100. Tablets containing citric acid released drug at a slower rate as a result of the suppression of polymer ionization due to a decrease in the micro-environmental pH of the tablet. The drug release profiles of the extruded tablets were found to fit both diffusion and surface erosion models (Bruce et al., 2005).

Theophylline or chlorpheniramine maleate pellets were coated with an aqueous ethylcellulose dispersion, Aquacoat. The influence of the plasticization time, curing conditions, storage time, and core properties on the drug release were investigated. The plasticization time (time between plasticizer addition to the polymer dispersion and the spraying process) did not affect the drug release, when the water-soluble plasticizer triethyl citrate was used, because of its rapid uptake by the colloidal polymer particles. In contrast, with the water-insoluble plasticizer acetyltributyl citrate, plasticization time (½ h vs 24 h) influenced the drug release, the longer plasticization time resulted in a slower drug release because of a more complete plasticizer uptake prior to the coating step. However, a thermal aftertreatment of the coated pellets at elevated temperatures (curing step) eliminated the effect of the plasticization time with acetyltributyl citrate. In general, curing reduced the drug release and resulted in stable drug release profiles. The time period between the coating and the curing step was not critical when the pellets were cured for a longer time. The structure of the pellet core (high dose matrix vs low dose layered pellet) strongly affected the drug release. A slow, zero-order drug release was obtained with high dose theophylline pellets, while a more rapid, first-order release pattern was obtained with low dose theophylline-layered nonpareil pellets (Wesseling & Bodmeier, 2001).

A pharmaceutical paste composition comprising the active ingredient such as an additive substance a control release agent and suitable carrier was patented (Odidi I. & Odidi A., 2009). The composition may be filled into a capsule or other dispensing device. Oily, waxy, or fatty substances were applied as plasticizers. Other invention relates to an oral pulse release comprising a polymer micromatrix, a first active ingredient distributed substantially uniformly within polymer micromatrix and a second active ingredient deposited on the surface of the polymer matrix (Gadre et al., 2006).

3.6 Drug release from plasticized polyesters

Films from poly(L-lactide) and poly(lactide-co-glycolide) were plasticized with polyethylene glycol. The plasticizer accelerated the degradation of polyester, its effect on the beginning of the release of the contained heparin significantly differed in dependence on the parameters of the polymer. In the homopolymer it decreased the burst-effect and accelerated the drug release in the phase controlled by diffusion, in copolymer the plasticizer did not exert a significant effect on the kinetics of release. The differences were explained by the influence of plasticizer on polymer hydrophilicity and crystallinity (Tan et al., 2004).

Thin films of a thickness of 40 μm from poly(lactide-co-glycolide) containing 10 % paclitaxel were plasticized with polyethylene glycols M_w 8,000 and 35,000 in various concentrations. The plasticizer with a lower molecular mass exerted a great influence on a more rapid release of paclitaxel. Polyethylene glycol was phase separated from copolymer (Steele et al., 2011).

The active ingredients were demonstrated to act as plasticizers of the polymer. The kinetics of dissolution in phosphate buffer of pH 7.4 and apparent diffusion coefficient including mathematical analysis of data resulted in the expression of the quantitative relationship between the diffusivity of drugs and the initial composition of medicinal substances with a possibility of prediction of the effect of thickness of the membrane and its composition on the kinetics of drug release (Siepmann et al., 2006).

Oligoester carrier compound of the equimolar ratio of glycolic acid and lactic acid branched with dipentaerythritol was synthesised by polycondensation (Snejdrova, E. & Dittrich, M., 2011). Table 3 shows the basic characteristic of the carrier.

Reactants proportion [weight %]			M_W [g/mol]	T_g [°C]
Lactic acid 47.5	Glycolic acid 47.5	Dipentaerythritol 5.0	2,300	16.3

Table 3. Relevant characteristics of oligoester carrier.

The carrier was plasticized using methyl salicylate in concentration of either 10 %, or 20 %, or 30 %. The increase of methyl salicylate concentration in the oligoester matrices influences the aciclovir release *in vitro* in the obvious manner. At 10 % concentration of the plasticizer the 90 % portion of active substance released was achieved after 15 days, at its 20 % concentration after 9 days , and at 30 % concentration after three days (Fig. 1).

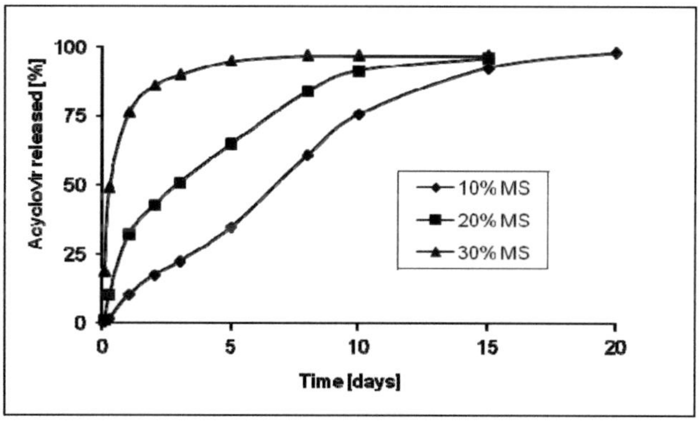

Fig. 1. Acyclovir release from oligoester carrier plasticized by methyl salicylate (MS).

Ethyl salicylate as more hydrophobic plasticizer in comparison to methyl salicylate was used in various concentrations for plasticizing of the oligoester carrier. It influences the aciclovir release kinetics in the more complicated way. The partition coefficient matrix/dissolution medium for drug is in the consequence a character of plasticizer and aqueous medium influx changed. During the drug release process small portion of plasticizer is separated as liquid heterophase containing aciclovir in the swelled matrices. The viscosity of matrices is dominant factor in the initial phase of drug release (Fig. 2).

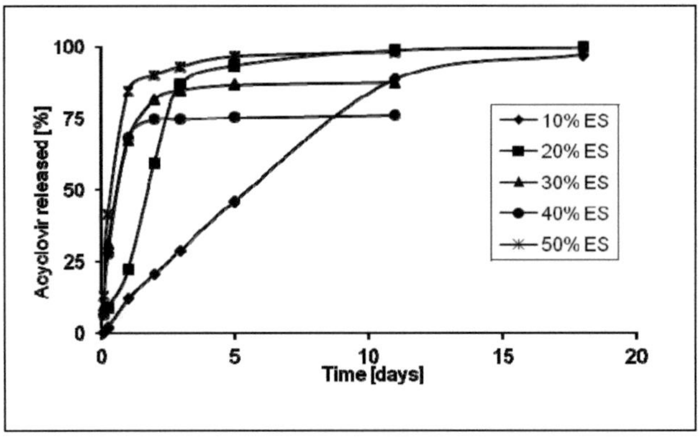

Fig. 2. Acyclovir release from oligoester carrier plasticized by ethyl salicylate (ES).

The influence of triethyl citrate concentration on acyclovir release from the branched oligoester carrier is shown in Fig. 3. Triethyl citrate differs from above mentioned plasticizers, methyl salicylate and ethyl salicylate, by its higher solubility in aqueous medium. The value of triethyl citrate solubility is 6.5 %, whilst for salicylic acid esters it is under 0.1 %. The accleration of dissolution process after 10 % triethyl citrate addition is expected situation. The hydrophilisation of matrices and their higher swelling is possible explanation for this behaviour. Opposite relation between concentration of triethyl citrate and velocity of drug release is reached at higher plasticizer concentrations. The possible hypothesis of this atypical situation is based on rapid collapse of matrices structure and formation of supramolecular structure based on more dense random coil conformation of molecules.

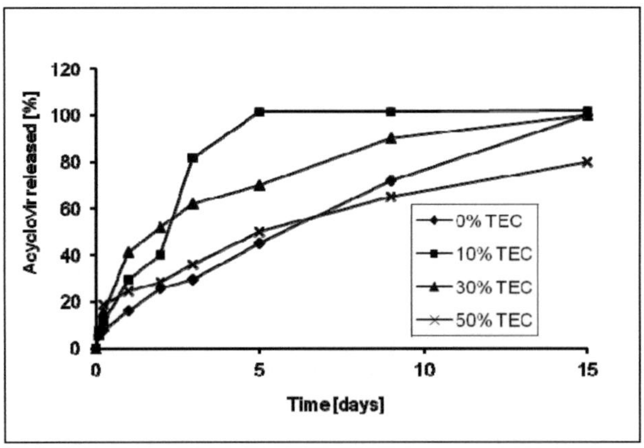

Fig. 3. Effect of triethyl citrate (TEC) on acyclovir release from branched oligoester carrier.

4. Conclusion

The "tailor made method" of selection the proper polymer for drug formulation has its limitations due to the demanding registration procedure of new polymers. Blending of two polymers or mixing of polymers with plasticizers and other additives are relatively simple and promising methods providing new biomaterials used in drug delivery with advantageous properties. Aliphatic polyesters are the most frequently used polymers in drug delivery systems. Star-like copolymers of hydroxy acids with polyhydric alcohols such as dipentaerythritol are particularly interesting. They possess good mechanical properties, however the brittleness is their major drawback for many applications. This is the reason for their blending with common and even non-traditional plasticizers. Viscosity of these drug carriers must be sufficiently low for good workability or in order their application via an injection needle or trocar applicator. The plasticizer type and concentration influence the whole profile of drug release.

5. References

Bader, H. & Göritz, D. (1994). Investigation on high amylose corn starch films. Part 2: Water vapor sorption. *Starch-Stärke,* Vol. 46, No.7, pp. 249–252, ISSN 1526-379X

Baiardo, M., Frisoni, G., Scandola, M., Rimelen, M., Lips, D. & Ruffieux, K. (2003). Thermal and mechanical properties of plasticized poly(lactic acid). *Journal of Applied Polymer Sciences,* Vol.90, No.7, (November 2003), pp. 1731-1738, ISSN 0021-8995

Baier, R.E. (1972). The role of surface energy in thrombogenesis. *Bulletin of the New York Academy of Medicine,* Vol.48, No.2, (February 1972), pp. 257-272, ISSN 0892-4503

Baltacioğlu, H. & Balköse D. (1999). Effect of zinc stearate and/or epoxidized soybean oil on gelation and thermal stability of PVC-DOP plastigels. *Journal of Applied Polymer Science,* Vol.74, No.10, (December 1999), pp. 2488–98, ISSN 0021-8995

Blasi, P., D'Souza, S.S., Selmin, F. & DeLuca, P. (2005). Plasticizing effect of water on poly(lactide-co-glycolide). *Journal of Controlled Release,* Vol. 108, No. 1, (November 2005), pp. 1 - 9, ISSN 0168-3659

Bodmeier, R. & Paeratakul, O. (1997). Plasticizer uptake by aqueous colloidal polymer dispersions used for the coating of solid dosage forms, *International Journal of Pharmaceutics,* Vol.152, No.1, (June 1997), pp. 17- 26, ISSN 0378-5173 177

Bourtoom, T. (2008). Plasticizer effect on the properties of biodegradable blend film from rice starch-chitosan. *Songklanakarin Journal of Sciences and Technology,* Vol.30 (Suppl.1), April, pp. 149-165, ISSN 0125-3395

Bruce, L.D, Shah, N.H., A., Malick, W., Infeld, M.H. & McGinity, J.W. (2005). Properties of hot-melt extruded tablet formulations for the colonic delivery of 5-aminosalicylic acid. *European Journal of Pharmaceutics and Biopharmaceutics,* Vol.59, No.1, (January 2005), pp. 85–97, ISSN 0939-6411

Carvalho, A.J.F., Zambon, M.D., Curvelo, A.A.S. & Gandini, A. (2003). Size exclusion chromatography characterisation of thermoplastic starch composites. 1. Influence of plasticizer and fibre content. *Polymer Degradation and Stability,* Vol.79, No. 1, (January 2003), pp. 133 - 138, ISSN 0141-3910

Chamarthy, S.P. & Pinal, R. (2007). Moisture induced antiplasticization in microcrystalline cellulose compacts, *Tablets & Capsules* , Vol.5, (July 2007), pp. 22–33, ISSN 1938-9159

Chamarthy, S.P. & Pinal, R. (2008). Plasticizer concentration and the performance of a diffusion-controlled polymeric drug delivery system. *Colloids and Surfaces A, Physicochemical Eneneering Aspects,* Vol.331, No.1-2, (December 2008) pp. 25–30, ISSN 0927-7757

Chaudhary, D.S., Adhikari, B.P. & Kasapis, S. (2011). Glass-transition behaviour of plasticized starch biopolymer system–A modified Gordon-Taylor approach. *Food Hydrocolloids,* Vol. 25, No.1, (January 2011), pp. 114 – 121, ISSN 0268-005X

Clayton, G.D. & Clayton, F.E. (Editors) (1993-1994). *Toxicology* 4th ed., p. 3042, ISBN: 0471547247 / 0-471-54724-7, New York, NY, John Wiley & Sons Inc.

Coyle, M., Martin, S.J. & McBrierty, V.J. (1996). Dynamics of water molecules in polymers. *Journal of Molecular Liquids,* Vol.69, (July 1996), pp. 95-116, ISSN 0167-7322

Epure, V., Griffon, E., Pollet, E. & Avérous, L. (2011). Structure and properties of glycerol-plasticized chitosan obtained by mechanical kneading. *Carbohydrate Polymers,* Vol.83, No.2, (January 2011), pp. 947-950, ISSN 0144-8617

European Medicines Agency (2009). Impurities: Guideline for residual solvents, Step 5, CH Topic Q3C (R4), pp. 7-22, London, United Kingdom

Fedorko, P., Djurado, D., Trznadel M., Dufour B., Rannou P. & Travers, J.P. (2003). Insulator–metal transition in polyaniline induced by plasticizers. *Synthetic Metals,* Vol.135–136, (April 2003), pp. 327–8, ISSN 0379-6779

Foldes E. (1998). Study of the effects influencing additive migration in polymers. *Die Angewandte Makromolekulare Chemie,* Vol. 261-262, No.1, (December 1998), pp. 65–76, ISSN 1522-9505

Gadre, A., Benmuvhar, M.R., Cheng, B.K.M., Gupta, V. & Herman, C.J. (2006). Matrix-based pulse release pharmaceutical formulation. Pat. PCT/US2006/010279 (October 12, 2006)

Hüettenrauch, R. & Fricke, S. (1984). Effect of glycerol on the helical conformation of gelatin. *Pharmazie,* Vol.39, No.8, (August 1984), pp. 500-501, ISSN 0031-7144

Jaeger, R.J. & Rubin, R.J. (1970). Plasticizers from plastic devices: extraction, metabolism and accumulation by biological systems. *Science,* Vol.170, No.460 (October 1970), ISSN 1095-9203

Johnson, W. Jr. (2002). Final report on the safety assessment of acetyl triethyl citrate, acetyl tributyl citrate, acetyl trihexyl citrate, and acetyl trioctyl citrate. *International Journal of Toxicology,* Vol.21, No.2 Supplement, (October 2002), pp. 1 – 17, ISSN 1093 -5818

Kambia, K., Dine, T., Azar, R., Gressier, B., Luyckx, M. & Brunet, C. (2001). Comparative study of leachability of di(2-ethylhexyl) phthalate and tri(2-ethylhexyl) trimellitate from haemodialysis tubing. *International Journal of Pharmaceutics,* Vol.229, No.1–2, (October 2001), pp. 139–146, ISSN 0378-5173 177

Kambia, K., Dine, T., Gressier, B., Bah, S., Germe, A.F., Luyckx, M., Brunet, C., Michaud, L. & Gottrand, F. (2003). Evaluation of childhood exposure to di(2-ethylhexyl) phthalate from perfusion kits during long-term parenteral nutrition. *International Journal of Pharmaceutics,* Vol.262, No.1–2, (August 2003), pp. 83–91, ISSN 0378-5173 177

Khandare, J. & Haag, R. (2010). Pharmaceutically Used Polymers: Principles, Structures, and Applications of Pharmaceutical Delivery Systems, In: *Drug Delivery: Handbook of*

Experimental Pharmacology 197, M. Schäfer-Korting, (Ed.), 221-250, Springer-Verlag Berlin, ISBN 978-3-642-00476-6, Heidelberg, Germany

Kim, K.S., Chin, I.J., Yoon, J.S., Choi, H.J., Lee, D.C.C Lee, K.H. ((2001). Crystallization behavior and mechanical properties of poly(etylene oxide)/poly(L-lactide/poly(vinyl acetate) blends. *Journal of Applied Polymer Sciences*, Vol.82, No.14, (December 2001), pp. 3618-3626, ISSN 0021-8995

Kissel, T., Brich, Z., Bantle, S., Lancranjan, I., Nimmerfall, F. & Vit, P. (1991). Parenteral depot-systems on the basis of biodegradable polyesters. *Journal of Controlled Release*, Vol.16, No.1-2, (June-July 1991), pp. 27–42, ISSN 0168-3659

Kowalczyk, D. & Baraniak, B. (2011). Effect of plasticizers, pH and heating of film-forming solution on the properties of pea protein isolate films. *Journal of Food Engineering*, Vol.105, No.2, (May 2011), pp. 295-305, ISSN 0260-8774

Kozlov, P.V. & Papkov, S.P. (1982). *Fiziko-chimicheskiye osnovy plastifikacii polimerow (The fundamentals of polymers plasticization)*, Chimiya, Moscow, USSR

Kulinski, Z., Piorkowska, E., Gadzinowska, K. & Stasiak, M. (2006). Plasticization of poly(L-lactide) with poly(propylene glycol). *Biomacromolecules*, Vol.7, No. 8, (July 2006), pp. 2128-2135, ISSN 1525-7797

Lakshmi, S. & Jayakrishnan, A. (1998). Migration resistant, blood compatible plasticized polyvinyl chloride for medical and related applications. *The International Journal of Artifician Organs*, Vol.22, No.3, (March 1998), pp. 222–229, ISSN 1724-6040

Lin, S.Y., Lee, Ch.J. & Lin, Y.Y. (1995). Drug-polymer interaction affecting the mechanical properties, adhesion strength and release kinetics of piroxicam-loaded Eudragit E films plasticized with different plasticizers. *Journal of Controlled Release*. Vol.33, No.3, (March 1995), pp. 375-381, ISSN 0168-3659

Liu, H., Chaudhary, D., Ingram, G. & John, J. (2011). Interaction of hydrophilic plasticizer molecules with amorphous starch biopolymer - An investigation into the glass transition and the water activity behavior. *Journal of Polymer Science, Part B, Polymer Physics*, Vol.49, No.14, (May 2011), pp. 1041-1049, ISSN 0887-6266

Liu, H., Xie, F., Yu, L., Chen, L. & Li, L. (2009). Thermal processing of starch-based polymers. *Progress in Polymer Science*, Vol.34, No.12, (December 2009), pp. 1348 - 1368, ISSN 0079-6700

Ljungberg, N., Colombini, D. & Vesslén, B. (2005). Plasticization of poly(lactic acid) with oligomeric malonate esteramides: dynamic mechanical and thermal film properties. *Journal of Applied Polymer Sciences*, Vol.96, No.4, (May 2005), pp. 992-1002, ISSN 0021-8995

Lowell Center for Sustainable Production (January 2011*). Phthalates and their alternatives: Health and Environmental Concerns*. University of Massachusetts Lowell , MA 01854, USA, pp. 1-23

Mali, S., Sakanaka, L.S., Yamashita, F. & Grossmann, M.V.E. (2005). Water sorption and mechanical properties of cassava starch films and their relation to plasticizing effect. *Carbohydrate Polymers*, Vol.60, No.3, (May 2005), pp. 283-289, ISSN 0144-8617

Martin, O. & Avérous, L. (2001). Poly(lactic acid): plasticization and properties of biodegradable multiphase systems. *Polymer*, Vol. 42, pp. 6209 - 6219, ISSN 0032-3861

Murphy, J. (2001). *Additives for plastics handbook*, Elsevier, ISBN 1856173704, New York, USA

Murray, W. (1867). On collodion dressings and applications. *British Medical Journal*, (April 27), pp. 495-496

Odidi, I. & Odidi, A. (2009). Pharmaceutical composition having a reduced abuse potential. US Pat. Appl. No. 20090232887 (September 17, 2009)

Pistel, K.F., Breitenbach, A., Zange-Volland, R. & Kissel, T. (2001). Brush-like branched biodegradable polyesters, part III: Protein release from microspheres of poly(vinyl alcohol)-graft-poly(D,L-lactic-co-glycolic acid). *Journal of Controlled Release*, Vol.73, No.1. (May 2001), pp. 7–20, ISSN 0168-3659

Rahman, M. & Brazel, Ch.S. (2004). The plasticizer market: an assessment of traditional plasticizers and research trends to meet new challenges. *Progress in Polymer Science*, Vol.29, No.12, (December 2004), pp. 1223–1248, ISSN 0079-6700

Rajewski, R.A & Haslam, J.L. (2008). Rapidly dissolving pharmaceutical compositions comprising pullulan. US Pat. Appl. No. 20080248102 (October 09, 2008)

Rasal, R.M., Janorkar, A.V. & Hirt, D.E. (2010). Poly(lactic acid) modifications. *Progress in Polymer Science*, Vol.35, No.3, (March 2010), pp. 338-356, ISSN 0079-6700

Ren, Z., Dong, L. & Yang, Y. (2006). Dynamic mechanical and thermal properties of plasticized poly(lactic acid). *Journal of Applied Polymer Sciences*, Vol.101, No.3, (August 2006), pp. 1583-1790, ISSN 0021-8995

Rosen, S.L. (1993). *Fundamental principles of polymeric materials*, John Wiley & Sons. ISBN 0-471-57525-9. New York, NY, USA.

Senichev, V.Y, & Tereshatov, V.V. (2004). Influence of plasticizers on the glass transition temperature of polymers, In: *Handbook of Plasticizers*, G. Wypych (Ed.), pp. 218–227, ChemTec Publishing, ISBN 1895198291, 9781895198294, New York, NY, USA

Sheth, M.M, Kumar, R.A., Davé, V., Gross, R.A. & McCarthy, S.P. (1997). Biodegradable polymer blends of poly(lactic acid) and poly(etylene glycol). *Journal of Applied Polymer Sciences*, Vol.96, No.1, (April 1997), pp. 149 - 1505, ISSN 0021-8995

Siepmann, F., Le Brun V, & Siepmann, J. (2006). Drugs acting as plasticizers in polymeric systems: A quantitative treatment. *Journal of Controlled Release*, Vol.115, No.3, (October 2006), pp. 298–306, ISSN 0168-3659

Sjoberg, P., Bondesson, U., Sedin, E. & Gustaffson, J. (1985). Exposure of newborn infants to plasticizers: plasma levels of di-(2-ethylhexyl) phthalate and mono-(2-ethylhexyl) phthalate during exchange transfusion. *Transfusion*, Vol.25, No.5, (September-October 1985), pp. 424–428, ISSN 1537-2995

Snejdrova, E. & Dittrich, M. (2011). Poly(α-hydroxyacids) as drug carriers. *Chemické Listy*, Vol.105, No.1, pp. 27-33, ISSN 0009-2770

Sobral, P.J.A., Menegalli, F.C., Hubinger, M.D. & Roques, M.A. (2001). Mechanical, water vapor barrier and thermal properties of gelatin based edible films. *Food Hydrocolloids*, Vol.15, No. 4-6, (July 2001), pp. 423-432, ISSN 0268-005X

Steele, T.W.J., Huang, Ch.L., Widjaja, E., Boey, F.Y.C., Loo, J.S.C. & Venkatraman, S.S. (2011). The effect of polyetylene glycol structure on paclitaxel drug release and mechanical properties of PLGA films. *Acta Biomaterialia*, Vol.7, No.5, (May 2011), pp. 1973-1983, ISSN 1742-7061

Tan, L.P., Venkatraman, S.S., Sung, P.F. & Wang, X.T. (2004). Effect of plasticization on heparin release from biodegradable matrices, *International Journal of Pharmaceutics,* Vol.283,No.1-2, (September 2004), pp. 89-96, ISSN 0378-5173

Tickner, J.A., Rossi, M., Haiama, N., Lappe, M. & Hunt, P. (1999). *The use of di(2-ethylhexyl) phthalate in PVC medical devices: exposure, toxicity and alternatives.* Center for Sustainable Production, Lowell, MA, 15. 7. 2011, Available from http://www.sustainableproduction.org/downloads/DEHP%20Full%20Text.pdf

Tsumura, Y., Ishimitsu, S., Kaihara, A., Yoshii, K. & Tonogai,Y. (2002) Phthalates, adipates, citrate and some of the other plasticizers detected in japanese foods: A survey. *Journal of Health Science,* Vol.48, No.5, (December 2002), pp. 493-502, ISSN 1347-5207

USP 35 (2011). The United States Pharmacopoeia 35. The United States Pharmacopoeial Convention. Mack Printing Co., Easton, WA, USA

Van De Wijdeven, G.G.P. (2010). Biodegradable material based on opened starch. US Pat. Appl. No. 20100015185 (January 04, 2010)

Wesseling, M. & Bodmeier, R. (2001). Influence of plasticization time, curing conditions, storage time, and core properties on the drug release from Aquacoat-coated pellets. *Pharmaceutical Development and Technology,* Vol.3, No.3, (August 2001), pp. 325-31. ISSN 1097-9867

Willet, J.L., Millard, M.M. & Jasberg, B.K. (1997). Extrusion of waxy maize starch: melt rheology and molecular weight degradation of amylopectin. *Polymer,* Vol.38, No.24, (November 1997), pp. 5983 – 5989, ISSN 0032-3861

Wilson, A.S. (1995). *Plasticizers principles and practice.* Cambrigde: The Institute of Materials. ISBN 0901716766, 9780901716767. London, England

World Health Organization (2003). *Concise International Chemical Assessment Document 52,* Diethyl Phthalate, WHO, Geneva, Switzerland

Wu, Ch. & McGinity, J.W (2001). Influence of Ibuprofen as a Solid-State Plasticizer in Eudragit® RS 30 D on the Physicochemical Properties of Coated Beads, In: *AAPS PharmSciTech,* 4. 7. 2011, Available from http://www.aapspharmscitech.org/view.asp?art=pt020424

Wu, Ch. & McGinity, J.W. (1999). Non-traditional plasticization of polymeric films. *International Journal of Pharmaceutics,* Vol.177, No.1, January, pp. 15-27, ISSN 0378-5173

Wu, Ch. & McGinty, J.W. (2000). Influence of relative humidity on the mechanical and drug release properties of theophylline pellets coated with an acrylic polymer containing methylparaben as an non-traditional plasticizer. *European Journal of Pharmaceutics and Biopharmaceutics,* Vol.50, No.2, (September 1999), pp. 277-284, ISSN 0939-6411

Yuksel, N., Baykara, M., Shirinzade, H. & Suzen, S. (2011). Investigation of triacetin effect on indomethacin release from poly(methyl methacrylate) microspheres: Evaluation of interactions using FT-IR and NMR spectroscopies. *International Journal of Pharmaceutics,* Vol.404, No.1, (February 2011), pp. 102–109, ISSN 0378-5173 177

Zhu, Y., Shah, N.H., Malick, A.W., Infeld, M.H. & McGinity, J.W. (2002). Solid-state plasticization of an acrylic polymer with chlorpheniramine maleate and triethyl citrate. *International Journal of Pharmaceutics, Vol.* 241, No.2, (July 2002) pp. 301–310, ISSN 0378-5173

Zondio Fiume, M. (2003). Final Report on the safety assessment of triacetin. *International Journal of Toxicology,* Vol.22, Supplement 2, (June 2003), pp. 1-10, ISSN 1091-5818

4

Pharmaceutical Applications of Plasticized Polymers

Eva Snejdrova and Milan Dittrich
Faculty of Pharmacy, Charles University in Prague
Czech Republic

1. Introduction

The modern pharmaceutical technology is not conceivable without plasticized polymers. The pharmaceutical applications of polymers range from packaging materials or auxiliary substances in conventional dosage forms to membranes or matrices modifying and controlling the drug release characteristics in therapeutic systems. Recently, a great variety of plasticized polymeric systems have been studied as microparticles, matrices, free membranes, membranes for transdermal systems and *in situ* forming implants. Not only the polymer itself but thorough choice of other ingredients of the polymer system is necessary for required quality of drug delivery system. The plasticizer has the unsubstitutable place in drug polymer system due to its primary role, i. e. improve the flexibility and processability of polymers by lowering the glass transition temperature. The supramolecular structure and relevant properties of specific plasticized polymer drug delivery systems has attracted the interest of many researchers.

2. Plasticized polymers used as coatings of the dosage forms

It is common practice to coat oral solid dosage forms with polymeric materials for several purposes, such as modifying drug release, affording gastroprotection, protecting from environmental agents, masking unpleasant taste or just enhancing product aesthetics. A film coating is defined as a thin and uniform polymer based-layer of about 20 to 100 μm in thickness, which is applied to the surface of substrates such as tablets, granules, powders, capsules, particles or pellets (Porter, 1995).

Polymers used in coating can be categorized into two types: (i) non-functional or conventional film coating polymers for immediate release coatings; which improve the appearance, the handling, prevent dusting or (ii) functional coating polymers, which can be used to modify the pharmaceutical function of the dosage forms, and include the delayed release dosage forms and sustained release (or extended or prolonged) dosage forms.

Coating films can be usually prepared from organic solvent-based polymers or aqueous solvent-based latex dispersions. However, the aqueous film coating have been favoured in coating technology because of its wide applications, low environmental concerns, efficiency of the process, and wide commercial availability of coating materials. The cellulose ethers and esters are the major group of polymers used in the film coating process. The pharmaceutical compositions of two kinds of coated beads were patented. The polymer coatings of the beads differed in water solubility. This pharmaceutical composition comprised phtalates, trimellitates, adipates, sebacates, glycols, polyethers and alkyl citrates as the plasticizers (Paborji & Flugel 2011).

Cellulose acetate phthalate (CAP), widely used in enteric film coating, is available as a white powder or also as a 30 % solid nanodispersion (pseudolatex) (Aquacoat® CPD). Polyvinyl acetate phthalate is less susceptible to hydrolysis, which minimizes or limits the content of free phthalic acid and other free acids.

Cellulose acetate trimellitate (CAT) has three carboxylic groups on the aromatic ring and dissolves at a pH of 5.5. The plasticizers as triacetin, acetylated monoglyceride, diethyl phthalate are recommended, and to obtain the best enteric coating results from aqueous processing, ammoniated solutions of CAT in water are the best choice.

Hydroxypropyl methylcellulose phthalate (HPMCP) characteristics, particularly at the pH where dissolution occurs, are determined by the degree of substitution of the three kinds of substituent groups (i.e. methoxy, hydroxypropoxy, and carboxybenzoyl). Valuable plasticizers include triacetin, acetylated monoglyceride and diethyl phthalate.

Acrylic polymers as a group of various synthetic polymers, methacrylic amino ester copolymers, have diverse functionality in film coating. They can be used for rapidly disintegrating coatings, taste and odour masking, coloured or transparent coatings, etc., and are available mainly as the Eudragit® products.

Methacrylic acid copolymers contain free carboxylic acid groups, and therefore can be used for enteric coating purposes by forming salts with alkalis. Methacrylic acid copolymers are soluble at pH values higher than 5.5. The addition of a plasticizer in these polymers is necessary. Methacrylic acid copolymers are available as various grades of Eudragit® products.

2.1 The role of the plasticizer in the film coating

The plasticizer plays the major role in the all process of the oral dosage form coating, and significantly influences the quality of the coatings, particularly the release of incorporated drugs.

2.1.1 Film formation facilitated by plasticizer

The film-formation mechanism from the coating dispersions is a complex process influenced mainly with the type and also with the quantity of plasticizer. The temperature below which a coating will not form a continuous, cohesive film is called the minimum film forming temperature; it is the minimum temperature above the T_g of the polymer. In order to ensure the optimum conditions for film formation, the really used coating temperature during the application process should be 10-20 °C above the minimum film forming temperature (Lehmann, 1992). Since plasticizer addition to film coating formulation reduces the T_g of the polymer, as a result the temperature required for film coating is reduced by plasticization (Zhu et al., 2002).

A sufficient amount of time should be allowed for the plasticizer uptake to ensure a homogeneous distribution of the plasticizer in the coating film. The time between the addition of the plasticizer to the polymer dispersion and the coating step is usually called as the plasticization time and presents a critical process variable, especially when water-insoluble plasticizers are used. Water insoluble plasticizers have to be emulsified in the aqueous phase of the polymer dispersions. During plasticization of the polymer dispersions, the plasticizer partitions into the colloidal polymer particles and softens them thus promoting particle deformation and coalescence into a film upon drying. Commercial suppliers of polymer dispersions often recommend a relatively short plasticization time of 30 - 60 minutes irrespective of the solubility of the plasticizer. Plasticizers are incorporated in the amorphous parts of polymers while the structure and size of any crystalline part remains unaffected (Fedorko et al., 2003). The thermal treatment following the application of the coat is known as curing. During this stage of coating, the coated dosage form is subjected to temperatures higher than the T_g, which facilitates uniform distribution of plasticizers. During curing of the coatings, plasticizers could also redistribute in the polymeric film.

Controlled released pharmaceutical formulations comprising pseudoephedrine embedded in core and coating shell made from mixture of two different polymers were approved. One of the used polymers was water soluble, the second one water insoluble. The water insoluble plasticizer ranged from 0.1 to 15.0 % by weight in this composition (Nair et al., 2010).

2.1.2 Mechanical properties of the coating improved by plasticizer

Good flexibility of the coating is essential for insuring the resistance of the coating to the mechanical impact force expected during the coating process, and in the application place of the coated dosage form (particularly in the gastrointestinal tract). However, pure polymer coatings are often brittle. An addition of an external plasticizer to the polymeric networks will increase the flexibility of the coatings (Siepmann et al., 1999).

The affirmative effect of the plasticizers on mechanical properties of the coating is caused by reducing of the cohesive intermolecular forces along the polymer chains. The flexibility is enhanced by increasing strain elongation and decreasing tensile strength and elastic modulus of the polymer (Gutierrez-Rocca & McGinity, 1994).

For example, triethyl citrate, triacetin and acetyltriethyl citrate used as a water-soluble plasticizer of the hydroxypropylmethylcellulose acetate succinate in press-coated tablets for colon targeting formulation ensure the plastic deformation property of the outer shell due to some interaction between hydroxypropylmethylcellulose acetate succinate and triethyl citrate (Fukui et al., 2001). Plasticized polymeric coating with dispersed bioactive anti-inflammatory and anti-thrombogenic drugs was applied for stents (Udipi & Cheng, 2004).

2.1.3 Adhesion of coating influenced by plasticizer

Good adhesion between a polymer and the surface of a solid is a major prerequisite for the film coating of pharmaceutical dosage forms. Loss of adhesion may reduce the functions that the film-coating provides to the solid substrate, particularly the mechanical protection that the coating ensures to the solid substrate and the drug release modified by coating.

There are two substantial forces which influence the polymer adhesion: the strength of the interfacial bonds between the polymeric film and the surface of the solid, and the internal stresses within the film coating. The interfacial bonding between the tablet surface and polymer coating is made primarily with hydrogen bonds, to a lesser extent with dipole-dipole and dipole-induced dipole interactions. The addition of plasticizing agents to coating formulations generally decreases the internal stress in the film by decreasing both the elastic modulus and the glass transition temperature of the film coating (Johnson & Hathaway, 1991; Rowe, 1981, 1992).

The measured force of adhesion, therefore, would be expected to increase with increased plasticization of the polymer. The contradictory results may be due to the interaction of the plasticizing agent with the polar groups of the polymer within the film structure (Felton & McGinity, 1999).

The adhesive property of Eudragit films is effected with organic esters used as plasticizers (triacetin, diethyl phthalate, dibutyl phthalate and tributyl citrate were tested). The molecular weight and solubility parameters of the plasticizer seemed to play an important role in changing the adhesive strength of the Eudragit films. The adhesion of Eudragit films is markedly increased when the concentration of the plasticizer is greater than 25%. Tributyl citrate may be the best choice of the plasticizer for the Eudragit film, particularly for the Eudragit E film.

The molecular weight and solubility parameters of the plasticizer seemed to play an important role in changing the adhesive strength of the Eudragit films (Lina et al., 2000).

The water soluble plasticizers, triethyl citrate and polyethylene glycol 6000, lower the Tg of the films to a greater extent than the hydrophobic plasticizers, tributyl citrate and dibutyl sebacate, and the coatings with the hydrophilic plasticizers exhibit stronger adhesion (Entwistle & Rowe, 1979; Felton & McGinity, 1997).

The study of the influence of plasticizers on the adhesion of the acrylic polymer to either hydrophilic or hydrophobic tablet compacts revealed that increasing tablet hydrophobicity decreased adhesion of both cellulosic and acrylic polymers (Felton & McGinity, 1996, 1997; Rowe, 1977).

2.2 Factors negatively influencing the coating

Polymer-plasticizer incompatibility, plasticizer migration out of polymer (and particularly leaching of plasticizer) as well sticking and aging of the film coated dosage influence not only the mechanical properties, but also drug release (Amighi & Moes, 1996; Arwidsson et al., 1991; Bodmeier & Paeratakul, 1992; Skultety & Sims, 1987). For example, dibutyl sebacate unlike triethyl citrate was found to remain within the polymeric coating upon exposure to the dissolution media and thus mechanical properties of coatings during drug release are maintained. Leaching of triethyl citrate out of the film resulting in decreased mechanical resistances facilitates crack formation.

A blend of polymers can be used in the coating of solid dosage forms, and then the type of plasticizer must be chosen responsibly to ensure good quality of the coating. If the plasticizer has different affinities to the individual polymers forming the blend, the plasticizer redistribution within the polymeric system during coating occurs. As an example can serve pellets coated with blends of ethyl cellulose and Eudragit® L. Dibutyl sebacate used as plasticizer of this coating shows a higher affinity to ethyl cellulose than to Eudragit® L (Lecomte et. al., 2004).

3. Free membranes

A membrane can be in the form of a free film containing the active ingredient, or it can form the outer wall of the body which is its reservoir together with other excipients. The membrane has a defined thickness; it can be porous or compact, it can contain, in addition to the active ingredient, various excipients including plasticizers. In the case of dosage forms such as buccal and sublingual films, capsules intended for oral, rectal or vaginal administration, and inserts placed into cavities, their membranes and membrane systems possess the function of (1) the primary coating increasing stability and enabling administration, (2) separation from and protection against the influence of the biological environment after administration, or (3) controlled release.

Ethyl cellulose films of standard thickness were prepared by slow evaporation of the solvent from chloroform solutions of a 48.8% ethoxylated cellulose derivative and three plasticizers: classic dibutyl sebacate, vitamin D3 and vitamin E. The individual plasticizers were used in nine different concentrations up to 80 %. It has been demonstrated that primarily α-tocopherol (vitamin E) in concentrations of 40 to 50 % is an advantageous and effective plasticizer of ethyl cellulose from the standpoint of toxicity, ultimate strength, toughness, and Young modulus (Kangarlou et al., 2008). Ethyl cellulose of the identical parameters was employed in the preparation of films using the same method, diethyl sebacate in a 40% concentration being used as the plasticizer. Chloroform was evaporated using the standard very slow evaporation procedure. A suitable velocity of stress action was found for the most conclusive stress-strain deformation diagrams (Sogol & Ismaeil, 2011).

Thin films made of poly(lactide-co-glycolide) were prepared from a dichloromethane solution with additions of 15%, 25%, or 50% polyethylene glycol 8 000 and 35 000, yielding potential implants of 40 μm in thickness containing the antitumor drug paclitaxel. Raman spectrometry was used to demonstrate partial separation of the plasticizer (Steele et al., 2011).

Soft gelatin capsules were formed by a wall of varying thickness containing about 40 to 45 % of gelatin, which is plasticized with water 20 to 30 % and glycerol, or sorbitol 30 to 35 % (Marques et al., 2009). The plasticizer increased the flexibility of gelatin, with increasing concentration to 45 % it decreased the sensitivity to humidity, decreased the tendency to recrystallization, decreased tensile strength and the elastic modulus, increases elongation at break (Bergo & Sobral, 2007).

Protein films possess good mechanical properties, but they are very sensitive to atmospheric humidity. Their hygroscopic character can be decreased by adding a hydrophobic plasticizer. Esters derived from citric acid such as triethyl citrate, tributyl citrate, acetyl triethyl citrate and acetyl tributyl citrate were tested in various concentrations (Andreuccetti et al., 2009). Their limited miscibility with an aqueous solution of gelatin was solved by emulsification of plasticizers with soya lecithin. The tensile stress of the material was decreased with increasing concentrations of the plasticizer, permeability of water vapours was moderately increased.

Biodegradable films were prepared from a protein isolated from sunflower oil with additions of various polyalcohols, such as glycerol, ethylene glycol, propylene glycol, polyethylene glycols. The mixture was subjected to thermo-moulding at 150 °C using 150 MPa. Tensile strength of films was 6.3 to 9.6 MPa and elongation at break was 23 to 140 %. Permeability of water vapours was low; films were resistant against aqueous environment only in a limited degree, as migration of the plasticizer took place (Orliac et al., 2003). The plasticizers used were evaluated as suitable for protein films.

4. Polymeric membranes for transdermal system

Transdermal systems are based on the adhesion of the system containing the active ingredient onto the intact skin and the transfer of the dissolved active ingredient through the skin. The limiting factor for the velocity of transdermal absorption is most frequently the thin membrane or a polymeric matrix structure. To fulfil the necessary precondition of the barrier principle as the controlling system of transdermal absorption, the velocity of permeation through the membrane or matrix must be lower than the velocity of penetration into the skin structures. Plasticizers are of extraordinary importance as they ensure not only suitable mechanical properties, in particular flexibility of the system, but also in connection with a decreased value of T_g increase the diffusivity of polymers.

In the development of a suitable membrane for transdermal use, cellulose acetate casted from a chloroform solution was used. The polymer was plasticized with dibutyl phthalate, polyethylene glycol 600 and propylene glycol in a concentration of 40 %. The prepared membranes were characterized from the standpoint of mechanical properties, transfer of water vapours and permeability of diltiazem hydrochloride and indomethacin. Permeability of the active ingredient was the highest with the use of polyethylene glycol as the plasticizer, the lowest in the case of dibutyl phthalate (Rao & Diwan, 1997). The plasticizer of membranes from cellulose acetate was poly(caprolactone triol), and the pore forming agent was water (Meier et al., 2004). On the basis of testing the permeability of paracetamol, water was demonstrated to be a suitable porosigen with a possible regulation of porosity by its concentration, the plasticizer suitably acting as the modulator of permeation of the active ingredient. A combination of these two excipients yielded the membranes whose values of the coefficients of permeation of paracetamol ranged from 10^{-7} to 10^{-5} cm s^{-1}.

The matrix system from polyvinyl acetate containing polyethylene-2 oleyl ether as the enhancer of permeation and triethyl citrate as the plasticizer resulted in a significant increase in the bioavailability of the antihistamine drug triprolidine. The effect was demonstrated after the administration of the system to the abdominal rabbit skin (Shin & Choi, 2005). Roughness, mechanical and adhesive properties of skin patch were studied. An adhesive plaster was prepared from a blend of polyvinyl alcohol and polyvinylpyrrolidon with glycerol or polyethylene glycols 200 or 400 as the plasticizers. The adhesivity of the system was ascertained only in contact with wet surfaces. Glycerol in a concentration of 10 % did not influence the very smooth texture of the surface of the plaster. The effect of the plasticizer was increased with decreasing molecular mass (Gal & Nussinovitch, 2009).

5. Matrix polymeric systems

Matrix polymeric systems are heterogeneous systems composed of a polymer, plasticizers, other additives and the active ingredient, and they can be porous. They include primarily oral preparations with prolonged or retarded release of the active ingredient, or implants produced by extrusion or hot moulding or compression.

Extrusion is a process of conversion of raw materials to form a product of a uniform shape and density by forcing it through a nozzle under controlled conditions (Breitenbach, 2002). Plasticizers facilitate the manufacture and improve the parameters of preparations such as porosity, tortuosity, mechanical resistance, diffusivity, or compatibility of the components. Hot-melt extrusion, in which it is possible to process materials with higher T_g values, is advantageous.

Due to their little volatility, the plasticizers which are in the solid state in room temperature are advantageous. For plasticization of polyacrylate polymers, e.g. matrices from Eudragit RS PO, citric acid, in particular its monohydrate, turned out to be useful. Tensile strength and the elastic modulus were decreased, whereas elongation was increased. The plasticizing effect of citric acid achieved the plateau stage at its 25% concentration in polyacrylate, i.e. when achieving the solubility value of (Schilling et al., 2007). The continous process of extrusion and subsequent division of the extruded body into tablets was used in the formulation of enterosolvent preparations from the polyacrylate material Eudragit L100-55. Mixtures of two plasticizers, triethyl citrate and citric acid, were employed. Triethyl citrate markedly decreased the T_g value and thus decreased the temperatures in the individual zones of the extruder and the pressure during extrusion. The optimal concentration of citric acid was 17 %. 5-aminosalicylic acid was employed as the model active ingredient (Andrews et al., 2008).

Subcutaneous implants of a diameter of 1.5 mm and length of 18 mm weighing 40 mg containing thermolabile recombinant proteins and peptides were prepared from plasticized poly(lactide-co-glycolide) as the carrier and ethanol as the plasticizer (Mauriac & Marion, 2006). First, a 10% solution of the polymer in ethanol was prepared, followed by drying to achieve an ethanol concentration of 20 %. The polymer plasticized in such a way was mixed with the unplasticized one and the mixture was extruded at 75 °C. The following step was grinding of the extrudate at -5 to -10 °C with an ethanol content of 8 %. The active ingredient and other additives were added to the powder and the mixture was extruded at a temperature up to 70 °C. Another example of extremely volatile plasticizer is supercritical carbon dioxide (Lakshman, 2008). This compoud is added to facilitate processing of materials by extrusion method.

Starch is the biomaterial which has excellent parameters from the standpoint of availability, price and biocompatibility. In the presence of water and suitable polar hydrophilic compatible plasticizers during extrusions at temperatures over 100 °C amylopectin flows and at the same time native starch granules partially melt. The material can have two markedly different values of T_g. Interactions between starch and plasticizers with strong hydrogen bonds (Pushpadass et al., 2008) were demonstrated in it. The use of glycerol in a concentration of 25 % and secondary plasticizers such as stearic acid, sucrose or urea in markedly lower concentrations are advantageous. It thus yields a flexible material suitable as a biodegradable drug carrier.

The matrix system containing a plasticized polymer can be also prepared by coating the pellets with a plasticized polymer and their subsequent compression to form tablets (Abdul et al., 2010). The use of a plasticizer does not require a curing step, which is heating after preparation. Pellets without a plasticizer are very brittle and break on compression. After an addition of 10 % triethyl citrate to the polymer Kollicoat SR30D (aqueous colloidal dispersion of polyvinyl acetate) the flexibility of the material was dramatically improved (Savicki & Lunio, 2005).

6. *In situ* forming implants

The classic medicated implants are the bodies of the solid state and of a defined shape, which are administered by an invasive surgical intervention into the muscular or other tissue. Their implantation requires local anaesthesia and a surgical intervention. *In situ* forming implants (ISFI) are liquid systems with a relatively low viscosity administered by an injection needle or a trocar containing a substance or several substances, which in the site of administration due to the biological environment spontaneously change its or their properties, in particular the mechanical ones. An important parameter is easy elimination capacity of systems based either on their biodegradability or slow dissolution (Hatefi & Amsden, 2002). The systems contain an active ingredient which is released from such implants in a long-term period, in the order of weeks to months. It thus enhances the effect of active ingredients, with decreased fluctuation of plasma concentrations toxicity is decreased, with decreased frequency of dosing the compliance of the patient is increased. Besides easy administration, an advantage of this relatively new dosage form is its simple manufacture. The systems find their use in human medicine, veterinary medicine (Winzenburg et al., 2004), and tissue engineering (Quaglia, 2008).

ISFI are suitable for both local and systemic administration of substances with antimicrobial activity, antitumor substances, hormones, substances interacting with the immunity system, growth factors, etc. The importance of this mode of administration will be increased with the introduction of other medicinal substances of the protein type, as the structure of the implant can protect these substances against their enzymatic decomposition.

The first system is named Atrigel Technology® and was patented in 1990 (Dunn et al., 1990; Warren et al., 2009). It is based on the administration of solutions of biodegradable polymers into the soft tissue. After administration, the water-soluble solvent is distributed into the surrounding tissue and the polymer is precipitated due to a backflow of aqueous solutions from the biological environment. The development of implants was studied *in vivo* using a non-invasive ultrasound imaging technique. The release of the active ingredient from implants has been demonstrated to be influenced not only by the composition of the surrounding pathologically changed tissue, but also by the mechanical conditions in it (Patel et al., 2010), and the process of precipitation of the polymer is considerably influenced by its molecular mass (Solorio et al., 2010). A system with continuous release of the immunoenhancer thymosin alpha 1 was constructed by dispersion of chitosan or albumin microparticles with this substance in the poly(lactide-co-glycolide) matrices dissolved in N-methyl-2-pyrrolidone. This achieved thymosin release for a period of 15 to 30 days (Liu et al., 2010).

The problem of the system Atrigel Technology® is the toxicity of solvents and a sudden initial release of a substantial part of the total dose of the contained active ingredient. In spite of it, in praxis there are antimicrobial and hormonal preparations based on this principle. The system based on the principle of rapid precipitation of the solutions of the biotechnological copolymer PHB/PHC in different solvents was employed to formulate ISFI of the film type acting preventively against adhesions of the tissues as undesirable phenomena in post-surgical applications (Dai et al., 2009).

Other types of ISFI are three-block copolymers of ABA or BAB types based on the gelation of their solutions after administration due to increased temperatures. Oligomers composed of polyethylene glycol blocks and the blocks of polyesters of aliphatic hydroxy acids are advantageous (Quiao et al, 2007; Tang & Singh, 2009). An advantage of the system ReGel® is the absence of toxic organic solvents and a solubilizing potential of the block copolymer. In pharmacotherapeutic praxis the system ReGel® with paclitaxel called OncoGel® is employed (Matthes et al., 2007;). For some active ingredients, some polymers and some modes of administration a too intensive burst effect, changes in the velocity of release of the active ingredient, or irritability of the polymer, which in the systems is used in higher concentrations, can occur (Packhaeuser et al., 2004).

Instead of polymer solutions in hydrophilic solvents, hydrophobic solvents of lower concentrations than the polymer concentration can be employed. They are thus the plasticized polymers. The behaviour of the system after its administration into the tissue is different, the system is not distributed into the environment. The polymer and possibly the plasticizer are subject to biodegradation, the mechanism of the release of the active ingredient is due to the enzymatic or hydrolytic destruction of the implant. A sufficiently low viscosity can be achieved, besides the use of the plasticizer, by increasing the temperature of the applied system. The maximal painless temperature in humans is stated as 53 °C, the maximal tolerable temperature without necrotic changes is 60 °C (Liu & Wilson, 1998).

The flowable composition relates to a sustained released delivery system with risperidone was patented. It may be injected into the tissue whereupon it coagulates to become the solid or gel, monolitic implant (Dadey, 2010).

An in-situ-hardening paste, containing a biodagradable polymer and water soluble polymeric plasticizer was developed as delivery system for an active agent in the field of tissue regeneration. The hardened paste can be used as bone and cartilage replacement matrix (Hellebrand et. al., 2009).

The following Table 1 presents a survey of hydrophobic plasticizers suitable for ISFI.

Substance	Melting point	Boiling point	Solubility in water [%]
Benzyl alcohol	-15 °C	205 °C	4.0
Benzyl benzoate	70 °C	324 °C	1.5×10^{-5}
Ethyl heptanoate	-66 °C	189 °C	3.0×10^{-4}
Propylene carbonate	-55 °C	242 °C	17.5
Triacetin	3 °C	258 °C	7.0
Triethyl citrate	-55 °C	235 °C / 20 kPa	6.5

Table 1. Characteristics of selected plasticizers with limited miscibility with water.

ISFI of this type use the protected name Alzamer® Depot technology (Alza Corporation) and are intended for subcutaneous administration. Thanks to the hydrophobic plasticizer, the systems possess a lower speed of degradation with a smaller burst and a slower release of the active ingredient (Matschke et al., 2002). Biocompatibility of plasticizers is higher than in the case of hydrophilic solvents. The advantage of most plasticizers of this type is their low volatility. Proteins and peptides are not dissolved in the systems, their suspensions are chemically very stable (Solanki et al., 2010). For a sufficiently decreased miscibility of ISFI of this type with the surrounding tissue liquid it is necessary to have the solubility of plasticizers in water lower than 7 % (Brodbeck et al., 2000). In situ forming thin membranes were prepared by mixing poly(lactide-co-glycolide) with 10 % polysorbate 80 as the plasticizer (Koocheki et al., 2011).

7. Plasticized polymeric microparticles

It is possible to differentiate two types of microparticles on the basis of their supramolecular texture. Microcapsules consist of a central part, which is called the core, and the peripheral part called the wall. The core contains the active ingredient; the wall is composed of the polymer. The microspheres are composed of a blend of a polymer and a suspended or dissolved active ingredient in the whole volume. The purpose of plasticization of polymers as the carriers of active ingredients in the microparticles is the modification of the kinetics of release either by changes in their hydrophilicity or lipophilicity, or a change in the T_g value, or a change in crystallinity. Release of active substances connected with these parameters of polymeric materials, and can be influenced by swelling, erosion, desorption, obstructive phenomena, osmotic phenomena, capillarity, etc.

Microspheres from the blends of hydroxypropyl methylcellulose and ethylcellulose with metoprolol were plasticized with diethyl phthalate. With increasing concentration of the active ingredient, the velocity of release was decreased. Microparticles plasticized in a 5% concentration of the hydrophobic excipient possessed the velocity of release identical with that of the non-plasticized microspheres, an increase of the concentration of the plasticizer from 15 % to 20 % resulted in a significant increase in the velocity of release (Somwanshi et al., 2011). Indomethacin-loaded microspheres from poly(methyl methacrylate) intended for oral administration were prepared. The period of release of the active ingredient from non-plasticized microparticles was 8 h. After addition of triacetin, the release lasted for 24 h. Fourier transform infrared and nuclear magnetic resonance spectra demonstrated an interaction of the hydrogen type bonds between indomethacin and the polymer. No interaction was demonstrated between triacetin and indomethacin or the polymer (Yuksel et al., 2011).

The liquid pharmaceutical formulations for oral administration with modified release of active principle, excluding amoxicillin was patented. Microcapsules prepared by microencapsulation of the active ingredient using proper polymer-plasticizer mixture were homogeneously dispersed into the external liquid phase (Castan et. al, 2011).

In the course of release, the microspheres from poly(DL-lactide) were plasticized with water from the dissolution medium. T_g values rapidly decreased. At temperatures higher than T_g a rapid erosion of microspheres occurred, whereas at lower temperatures the polymer did not change for the period of the experiment (Aso et al., 1994). The microspheres from poly(L-lactide) and poly(lactide-co-glycolide) were obtained. After preparation using a standard solvent evaporation technique after freeze and oven drying these microparticles contained up to 3 % of water. Residual water markedly decreases T_g values according to Gordon-Taylor relationship (Passerini & Craig, 2001).

The microspheres intended for target oriented drug distribution were prepared from poly(lactic-co-glycolic acid) containing the chemotherapeutic agent etoposide in various concentrations. Plasticization with tricaprin in concentrations of 25 % and 50 % significantly increases the velocity of etoposide release in comparison with the microspheres without a plasticizer (Schaefer & Singh, 2002). The microcapsules containing the active ingredients soluble in water were prepared by the o/o/o emulsion method under the extraction of the solvent. Peanut oil was employed in the middle oily phase of multiple emulsion (Elkharraz et al., 2011). This peanut oil plasticized the internal phase containing a solution of the active ingredient and poly(DL-lactide) or poly(lactide-co-glycolide) .

The velocity of release of the anticancer agent paclitaxel from poly(lactic-co-glycolic) microspheres was increased after addition of 30 % of isopropyl myristate, 70 % of the active ingredient was released within 3 weeks. After an increase in the concentration of the plasticizer to 50 % there was another increase in the velocity of the process. The plasticizer did not influence the course of degradation of polymers. Release of paclitaxel took place by the mechanism of diffusion from minimatrices (Sato et al., 1996). Analogous conclusions were published in a similar case of microspheres with etoposide (Schaefer & Singh, 2000).

In situ forming microparticle systems are based on the emulsification of the solution of the active ingredient and polymer in the outer oily or aqueous phase. After the application of the emulsion there occurs separation of the solvent to the biological environment and solidification of the system. Besides water-soluble solvents it is possible to use the more hydrophobic, in a limited degree water soluble ones, which act as plasticizers. Myotoxicity of the plasticizers of this type is lower; the following series of decreasing toxicity was found: benzylalcohol > triethyl citrate > triacetin > propylene carbonate > ethyl acetate. Myotoxicity of ethyl acetate was comparable with the isotonic sodium chloride (Rungseevijitprapa et al., 2008).

8. Bioadesive plasticized polymers

Plasticized polymers play an unsubstitutable role in the formulation of the bioadhesive drug delivery systems (BDDS). Bioadhesive polymers have been formulated into tablets, patches, or microparticles, typically as a matrix into which the drug is dispersed, or as a barrier through which the drug diffuses (Ahuja et al., 1997).

In most instances bioadhesive formulations are preferred over the conventional dosage forms, because bioadhesion allows the retention of the active substance in the place of application, or even absorption, and thus increases transmucosal fluxes. As mostly the hepatic first-pass metabolism is avoided, drug bioavailability may be enhanced. Target sites for bioadhesive drug delivery include the eye, GIT, oral cavity, nasal cavity, vagina, and cervix. The development of the adhesive dosage forms for controlled drug delivery to or via mucous membranes is of interest with regard to local drug therapy, as well the systemic administration of peptides and other drugs poorly absorbable from the gastrointestinal tract. There is an important difference between the technical adhesion and bioadhesion; it is the presence of water, which is necessary for bioadhesion but impedes most technical applications.

Similar to the plasticization mechanism also the mechanism of bioadhesion is usually analysed based on the polymer chains attractive and repulsive forces. Generally, bioadhesion is regarded as a two-step process. The first step is considered to be an interfacial phenomenon influenced by the surface energy effects and spreading process; a second step involves chain entanglement across a large distance, i.e. polymer chains show interdiffusion. Plasticizers can significantly influence both of these mentioned steps.

According to the wetting theory of bioadhesion (Smart, 2005), there is a significant effect of the viscosity of the plasticized polymer on adhesivity. A plasticized polymer due to a lower viscosity has a higher ability to spread onto a surface as a prerequisite for the development of adhesion.

The plasticizer reduces the aggregation process caused by the intermolecular attraction of the polymer, and it results in an increase in bioadhesiveness. The strength of bioadhesion should be sufficient, but not so sharp as to damage the biological tissue in the application site. The molecular weight, solubility parameter and concentration of the plasticizers used play an significant role. Further, a lower molecular weight or a higher concentration of plasticizers might lead to a greater plasticizing action (Qussi, & Suess, 2006).

Since the strength of adhesion is dependent on the number and type of interfacial interactions, different polymers and the way of their plasticization will exhibit different adhesive properties depending on both the chemical structure and physico-chemical properties of the polymer, as well the plasticizer used. Many approved pharmaceutical excipients, which are well known and widely used, possess bioadhesive properties and are the first choice candidates for the formulation of bioadhesive preparations particularly due to easier registration.

Bioadhesive materials are generally hydrophilic macromolecular compounds that contain numerous hydrogen bond forming groups, notably carboxyl, hydroxyl, amide and amine groups, and will hydrate and swell when placed in contact with an aqueous solution. Most often these materials need to hydrate to become adhesive but overhydration usually results in the formation of a slippery mucilage and loss of the adhesive properties (Peppas & Sahlin,1996). The invention provides mucoadhesive bioerodible, water soluble carriers for ocular delivery of pharmaceuticals for either systemic or local therapy (Warren et al., 2008).

Plasticizer efficiency in bio/mucoadhesion is negatively influenced by inducing of the drug-polymer interactions. In this case, the drug in BDDS is not only the active ingredient but represents an antiplasticizing additive responsible for the lowering of the adhesion strength.

The triacetin-plasticized Eudragit E can serve as a film-forming material for the self-adhesive drug-loaded film for transdermal application of piroxicam. Piroxicam did not represent only a simple model drug, it acts as an additive by molecularly dispersing it in the Eudragit E film. Drug-polymer interaction occurring between piroxicam and the Eudragit E film might be responsible for a decrease in adhesion strength.

In order to improve the flexibility and adhesiveness of the drug-loaded Eudragit E film plasticized with triacetin, secondary plasticizer can be supplemented. Polyethylene glycol 200, propylene glycol, diethyl phthalate, and oleic acid can serve as the secondary plasticizer to improve adhesion (Lin, et al., 1991, 1995).

The adhesive properties were revealed in plasticized star-like branched terpolymers of dipentaeythritol, D,L-lactic acid and glycolic acids, where dipentaerythritol in concentration 3 %, 5 % , or 8 % as the branching agent was used in the synthesis by polycondenzation. The common plasticizer, triethyl citrate, as well the non-traditional plasticizers methyl salicylate, ethyl salicylate, hexyl salicylate, and ethyl pyruvate were used (Fig. 1). These multi-functional plasticizers can serve not only as plasticizers, but potentially pharmacodynamic efficient ingredients.

Presently, there is still no universal test method for bioadhesion measurement. Tensile testing systems are the most widely used in-vitro method for assessing the strength of the bioadhesive interactions. The instrumental variables such as contact force, contact time and speed of withdrawal of probe from the substrate can affect the results of the adhesion measurements.

The adhesiveness of the plasticized branched oligoesters was measured as the maximal force needed for the detachment of the adhesive material from the substrate, using the material testing machine Zwick/Roel T1-FR050TH.A1K (Snejdrova & Dittrich, 2009). The porcine stomach mucin gel served as a model substrate. All the plasticizers used provide high effectiveness in viscosity lowering, and thus a perfect spreading of the adhesive material on the contact surface as the prerequisite for good adhesion. Sufficient bioadhesion force was revealed in the wide range of dynamic viscosity (Fig. 2) (Snejdrova & Dittrich, 2008).

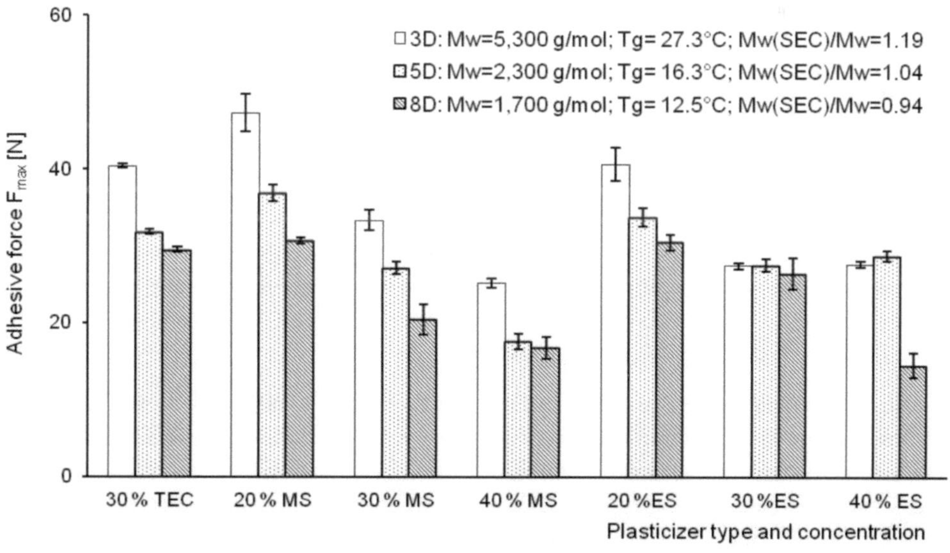

Fig. 1. Adhesiveness of branched oligoester carriers of the drug influenced by plasticizer type and concentration: triethyl citrate (TEC), methyl salicylate (MS), ethyl salicylate (ES).

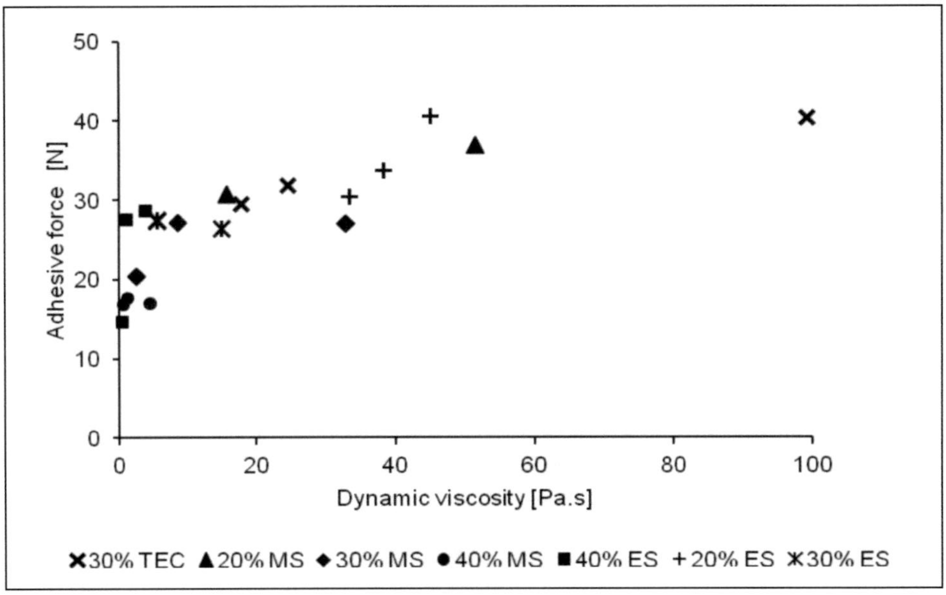

Fig. 2. Relation between viscosity and adhesivity of the branched oligoester carriers plasticized by various type and concentration of the plasticizers (triethyl citrate (TEC), methyl salicylate (MS), ethyl salicylate (ES).

9. Conclusion

The plasticization of polymers used in pharmaceutical technology can solve a lot of problems during the dosage forms formulation and can improve the quality of the final polymeric drug delivery system. The processing disadvantages can be thus overcome, or even a new technology can be enabled. The products of new quality are obtained by film coating of tablet with the thin layer of a plasticized polymer. The conventional film coating polymers mask the unpleasant organoleptic qualities of drug incorporated within the solid dosage form and protect the active drug substance from exposure to light, atmospheric moisture and oxygen. The functional coating polymers are used to modify the drug release profile, including the delayed release and sustained release. The membranes plasticized with suitable additive reaches the optimal parameters of permeability combined with the required drug release profile. Thanks to significant decrease of viscosity using the plasticizers, the *in situ* forming implants can be formulated and administered by an injection needle or a trocar. Multiparticulate dosage forms after incorporation of the proper plasticizer dispose of increased bioavailability.

Plasticized polymers play the unsubstitutable role in the formulation of the bioadhesive drug delivery systems. The adhesive properties were revealed in plasticized star-like branched terpolymers of dipentaeythritol, D,L-lactic acid and glycolic acid. The multifunctional plasticizers methyl salicylate, ethyl salicylate, hexyl salicylate, and ethyl pyruvate can serve not only as plasticizers, but potentially also as pharmacodynamic efficient ingredients. In accordance with the spreading theory of bioadhesion, the decrease in viscosity resulted in improving of the bioadhesivity after addition of these plasticizers.

Pharmaceutically used polymer	Plasticizer	Applied as
Cellulose nitrate („collodion”)	castor oil	historically the first plasticized polymer used as a medicinal preparation (wounds covering)
PVC	di(2-ethylhexyl) phthalate, diisononyl phthalate, diisodecyl phthalate, epoxidized triglyceride, vegetable oils from soybean oil, linseed oil, castor oil, sunflower oil, fatty acid esters	medical devices (bags, catheters, gloves, intravenous fluid containers, blood bags, medical tubings)
Ethyl cellulose	dibutyl sebacate	coatings free membranes
Cellulose acetate	dibutyl phthalate, PEG 600, propylene glycol, poly(caprolactone triol)	polymeric membranes for transdermal system
Blends of hydroxypropyl methylcellulose and ethylcellulose	diethyl phthalate	microparticles
Cellulose acetate phthalate Cellulose acetate trimellitate Hydroxypropyl methylcellulose phthalate Polyvinyl acetate phthalate Shellac	triacetin, acetylated monoglyceride, diethyl phthalate	enteric or colonic drug delivery
Hydroxypropyl methylcellulose acetate succinate	triethyl citrate, triacetin, acetyltriethyl citrate	press-coated tablets for colon targeting
Blends of ethyl cellulose and Eudragit® L	dibutyl sebacate	enteric film coatings
Chitosan salts (chloride, lactate, gluconate)	glycerol, ethylenglycol, propylenglycol, PEGs	free membrane
Blend of native rice starch and chitosan	sorbitol, glycerol, PEG 400	free membrane
Thermoplastic starch Amylose Cassava starch	polyhydric alcohols (glycerol, xylitol, sorbitol), secondary plasticizers (stearic acid, sucrose, urea)	polymeric matrices
Gelatin	glycerol, sorbitol, mannitol, sucrose, citric acid, tartaric acid, maleic acid, PEGs	free membranes

Pharmaceutically used polymer	**Plasticizer**	**Applied as**
Whey protein Sunflower protein	triethyl citrate, tributyl citrate, acetyl triethyl citrate, acetyl tributyl citrate, glycerol, ethylene glycol, propylene glycol, PEGs	particulate systems
Poly(lactic acid) Poly(lactide-co-glycolide)	oligoesters or low-molecular polyesters, polyesteramides, PEGs, polypropylene glycol, blend of triacetin and oligomeric poly(1,3-butanediol), active ingredients (ibuprofen, theophylline, salts of metoprolol and chlorpheniramine), ethanol, polysorbate 80, water, tricaprin, peanut oil, isopropyl myristate	free membranes polymeric matrices *in situ* forming systems microparticles
Star-like branched terpolymers of dipentaeythritol, D,L-lactic acid and glycolic acids	triethyl citrate, methyl salicylate, ethyl salicylate, hexyl salicylate, ethyl pyruvate	bioadhesive drug delivery systems
Copolymers of methacrylate esters ammoniated (Eudragit® E grades)	tributyl citrate, triacetin, PEG 200, secondary plasticizer (propylene glycol, diethyl phthalate, oleic acid)	moisture protection and odor/taste masking coatings bioadhesive drug delivery systems
Copolymers of ethyl acrylate and methyl methacrylate (Eudragit® RL and RS grades)	water (relative humidity), citric acid monohydrate, active ingredients (metoprolol tartrate, chlorpheniramine maleate), auxiliary compounds (surfactants, preservatives- methylparaben, solvents, cosolvents, desolvating and coacervating agents)	time-controlled drug release polymeric matrices
Poly(methyl methacrylate)	Triacetin, supercritical carbon dioxide, dibutyl phthalate	microparticles intraocular lenses hard contact lenses
Blend of polyvinyl alcohol and polyvinylpyrrolidon	glycerol, PEG 200 or 400	polymeric membranes for transdermal system
Kollicoat SR30D (aqueous colloidal dispersion of PVA)	triethyl citrate	polymeric matrices
Various non-soluble polymers	benzyl alcohol, benzyl benzoate, ethyl heptanoate, propylene carbonate, triacetin, triethyl citrate	*in situ* implants

Table 2. List of plasticized polymers reviewed in presented chapter.

10. References

Abdul, S., Chandewar, A.V. & Jaiswal, S.B. (2010). A flexible technology for modified-release drugs: Multiple-unit pellet system (MUPS). *Journal of Controlled Release,* Vol.147, No.1, (October 2010), pp. 2-16, ISSN 0168-3659

Ahuja, A., Khar, R. P. & Ali, J. (1997). Mucoadhesive Drug Delivery System. *Drug Development & Industrial Pharmacy,* Vol.23, No.5, (January 1997), pp. 489 - 515, ISSN 1520-5762

Amighi, K. & Moes, A. (1996). Influence of plasticizer concentration and storage conditions on the drug release rate from Eudragit® RS 30 D film-coated sustained-release theophylline pellets. *European Journal of Pharmaceutics and Biopharmaceutics,* Vol.42, No.1, pp. 29-35. ISSN 0939-6411

Andreucatti, C., Carvalho, R.A. & Grosso, C.R.F. (2009). Effect of hydrophobic plasticizers on functional properties of gelatin-based films. *Food Research International,* Vol.42, No.8, (October 1994), pp. 1113-1121, ISSN 0963-9969

Andrews, G.P., Jones, D.S., Diak, O.A., McCoy, C.P., Watts, A.B. & McGinity, J.W. (2008). The manufacture and characterization of hot-melt extruded enteric tablets. *European Journal of Pharmaceutics and Biopharmaceutics,* Vol.69, No.1, (May 2008), pp. 264-273, ISSN 0939-6411

Arwidsson, H., Hjelstuen, O., Ingason, D. & Graffner, C. (1991). Properties of ethyl cellulose films for extended release. Part 2. Influence of plasticizer content and coalescence conditions when using aqueous dispersions, *Acta Pharmaceutica Nordica,* Vol.3, No.2, pp. 65-70, ISSN 1100-1801

Aso, Y., Yoshioka, S., Po, L.W. & Terao, T. (1994). Effect of temperature on mechanisms of drug release and matrix degradation of poly(DL-lactide) microspheres. *Journal of Controlled Release,* Vol.31, No.1, (August 1994), pp. 33 - 39, ISSN 0168-3659

Bergo, P. & Sobral, P.J.A. (2007) Effects of plasticizer on physical properties of pigskin gelatin films. *Food Hydrocolloids,* Vol. 21, No.8, (December 2007), pp. 1285-1289, ISSN 0268.005X

Bodmeier, R. & Paeratakul, O. (1992). Leaching of water-soluble plasticizers from polymeric films prepared from aqueous colloidal polymer dispersions. *Drug Delivery and Industrial Pharmacy,* Vol.18, No.17, (January 1992), pp. 1865-1882, ISSN 1520-5762

Breitenbach, J. (2002). Melt extrusion: from process to drug delivery technology. *European Journal of Pharmaceutics and Biopharmaceutics,* Vol.54, No.2, (September 2002), pp. 107-117, ISSN 0939-6411

Brodbeck, K.V., Gaynor-Duarte, A.T. & Shen, T.I. (2004). Gel compositions and methods, US Pat. No. 61032000 (March 21, 2000)

Castan, C., Guimberteau, F. & Meyrueix, R. (2011). Oral pharmaceutical formulation in the form of aqueous suspension of microcapsules for modified release of amoxicillin. US Pat. No. 7,910, 133 B2 (May 25, 2006)

Dadey, E. (2010). Sustained delivery formulations of risperidone compounds. US Pat. Appl. No. 20100266655 (October 21, 2010)

Dai, Z.W., Zou, X.H. & Chen, G.Q. (2009). Poly(3-hydroxybutyrate-co-3-hydroxyhexanoate) as an injectable implant system for preventive of post-surgical tissue adhesion. *Biomaterials,* Vol.30, No.17, (June 2009), pp. 3075-3083, ISSN 0142-9612

Dunn, R.L., English, P., Cowsar, D.R. & Vanderbilt, P. (1990). Biodegradable in-situ forming implants and methods of producing the same. US Pat. No. 4938763 (June 03, 1990)

Fedorko, P., Djurado, D., Trznadel M., Dufour B., Rannou P. & Travers, J.P. (2003). Insulator–metal transition in polyaniline induced by plasticizers. *Synthetic Metals*, Vol.135–136, (April 2003), pp. 327–8, ISSN 0379-6779

Felton, L.A. & McGinity, J.W. (1997). Influence of plasticizers on the adhesive properties of an acrylic resin copolymer to hydrophilic and hydrophobic tablet compacts. *International Journal of Pharmaceutics*, Vol.154, No.2, (August 1997), pp. 167–178, ISSN 0378-5173

Felton, L.A. & McGinity, J.W. (1999). Adhesion of polymeric films to pharmaceutical solids. *European Journal of Pharmaceutics and Biopharmaceutics*, Vol.47, No.1, (January 1999), pp. 3–14, ISSN 0939-6411

Felton, L.A., McGinity, J.W. (1996). Influence of tablet hardness and hydrophobicity on the adhesive properties of an acrylic resin copolymer. *Pharmaceutical Development & Technology*, Vol.1, No.4, (January 1996). pp. 381–389, ISSN 1083-7450

Fukui, E., Miyamura, N., Yoneyama, T. & Kobayashi, M. (2001). Drug release from and mechanical properties of press-coated tablets with hydroxypropylmethylcellulose acetate succinate and plasticizers in the outer shell. *International Journal of Pharmaceutics*, Vol.217, No.1-2, (April 2001), pp. 33–43, ISSN 0378-5173

Gal, A. & Nussinovich, A. (2009). Plasticizers in the manufacture of novel skin-bioadhesive patches. *International Journal of Pharmaceutics*, Vol.370, No.1-2, (March 2009), pp. 103-109, ISSN 0378-5173

Gutierrez-Rocca, J.C & McGinity, J.W. (1994). Infuence of water soluble and insoluble plasticizers on the physical and mechanical properties of acrylic resin copolymers. *International Journal of Pharmaceutics*, Vol.103, No.3, (March 1994), pp. 293-301, ISSN 0378-5173

Hatefi, A. & Amsden, B. (2002). Biodegradable injectable in situ forming drug delivery systems. *Journal of Controlled Release*, Vol.80, No.1-3, (April 2002), pp. 9-28, ISSN 0168-3659

Hellebrand, K., Seidler, M., Schütz, A., Pompe, C. & Friess, W. (2009). In situ hardening paste, its manufacturing and use. US Pat. Appl. No. 20090048145 (February 19, 2009)

Johnson, K., Hathaway, R., Leung, P. & Franz, R. (1991). Effect of triacetin and polyethylene glycol 400 on some physical properties of hydroxypropyl methycellulose free films. *International Journal of Pharmaceutics*, Vol.73, No.3, (July 1991), pp. 197–208, ISSN 0378-5173

Kangarlou, S.; Haririan, I. & Gholipour, Y. (2008). Physico-mechanical analysis of free ethyl cellulose films comprised with novel plasticizers of vitamin resources. *International Journal of Pharmaceutics*, Vol.356, No.1-2, (May 2008), pp. 153-166, ISSN 0378-5173

Lakshman, J.P. (2008). Process for making pharmaceutical compositions with a transient plasticizer. US Pat. Appl. No. 20080280999 (November 13, 2008)

Lecomte, F., Siepmann, J., Walther, M., MacRae, R.J. & Bodmeier, R. (2004). Polymer blends used for the aqueous coating of solid dosage forms: importance of the type of plasticizer. *Journal of Controlled Release*, Vol. 99, No.1, (September 2004) pp. 1- 13, ISSN 0168-3659

Lehmann, K. (1992). *Microcapsules and Nanoparticles in Medicine and Pharmacy*, CRC Press, ISBN 0-8493-6986-X, Boca Raton, Florida

Lin, S.Y., Chau-Jen Lee, Ch.J. & Lin, Y.Y. (1995). Drug-polymer interaction affecting the mechanical properties, adhesion strength and release kinetics of piroxicam-loaded Eudragit E films plasticized with different plasticizers. *Journal of Controlled Release*, Vol.33, No.3, (March 1995), pp. 375-381, ISSN 0168-3659

Lin, S.Y., Lee, Ch.J. & Lin., Y.Y. (1991). The effect of plasticizers on compatibility, mechanical properties, and adhesion strength of drug-free Eudragit E films. *Pharmaceutical Research*, Vol.8, No.9, (September 1991), pp. 1137-1143, ISSN 1573-904X

Liu, F. & Wilson, B. C. (1998) *Hyperthermia and photodynamic therapy,* in: Tannock, I., Hill, R.P. *(eds.), Basic Science of Oncology,* McGraw-Hill, New York, 1998, pp. 443 - 453, ISBN 0071053166

Liu, Q., Zhang, H., Zhou, G., Xie, S., Zou, H., Yu, Y., Li, G., Sun, D., Zhang, G., Lu, Y. & Zhong, Y. (2010). In vitro and in vivo study of thymosin alpha 1 biodegradable in situ forming poly(lactide-co-glycolide) implants. *International Journal of Pharmaceutics*, 397, No.1-2, (September 2010), pp. 122-129, ISSN 0378-5173

Marques, M.R.C., Cole, E., Kruep, D., Gray, V., Murachanian, D., Brown, W.E. & Giancaspro, G.I. (2009). Liquid-filled gelatin capsules. *Pharmacopoeial Forum,* Vol. 35, No.4, pp. 1020-1041, USPC

Matschke, Ch., Isele, U., van Hoogevest, P. & Ffahr, A. (2002). Sustained injectables formed in situ and their potential use for veterinary products. *Journal of Controlled Release*, Vol.85, No.1-3, (December 2002), pp. 1-15, ISSN 0168-3659

Matthes, K., Mino-Knudson, D.V., Sahani, N., Holalkere, N., Fowers, K.D, Rathi, R. & Brugge, W.R. (2007). EUS guided injection of paclitaxel (OncoGel) provides therapeutic drug concentrations in the porcine pancreas. *Gastrontestinal Endoscopy*, Vol.65, No.3, (March 2007), pp. 448-453, ISSN 0016-5107

Mauriac, P. & Marion, P. (2006). Use of ethanol as plasticizer for preparing subcutaneous implants containing thermolabile active principles dispersed on a plga matrix. US Pat. Appl. No. 20060171987 (July 24, 2004)

Meier, M.M., Kanis, L.A., Solodi, V. (2004). Characterisation and drug-permeation profiles of microporous and dense cellulose acetate membranes: influence of plasticizer and pore forming agent. *International Journal of Pharmaceutics*, Vol.278, No.1, (June 2004), pp. 99 -110, ISSN 0378-5173

Nair, R., Rajheja, P., Wang, S. & Pillai, R.S. (2010). Pseudoephedrine pharmaceutical formulations. US Pat. Appl. No. 20100260842 (October 14, 2010)

Orliac, O., Rouilly, A., Silvestre, F. & Rigal, L. (2003). Effects of various plasticizers on the mechanical properties, water resistance and aging of thermo-moulded films made from sunflower proteins. *Industrial Crops and Products,* Vol.18, No.2, (September 2003), pp. 91-100, ISSN 0926-6690

Paborji, M. & Flugel, R.S. (2011). Pharmaceutical formulations for the treatment of overactive bladder. US Pat. Appl. No. 20110244051 (June 10, 2011)

Packenheuser, C.B., Schnieders, J., Oster, C.G. & Kissel, T. (2004). In situ forming parenteral drug delivery system: an overview. *European Journal of Pharmaceutics and Biopharmaceutics*, Vol.58, No.2, (September 2004), pp. 445-455, ISSN 0939-6411

Passerini, N. & Craig, D. Q. M. (2001). An investigation into the effects of residual water on the glass transition termperature of polylactide microspheres using modulated temperature DSC. *Journal of Controlled Release,* Vol.73, No.1, (August 1994), pp. 111-115, ISSN 0168-3659

Patel, R.B., Solorio, L., Wu, H., Krupka, T. & Exner, A.A. (2010). Effect of injection site on in situ implant formation and drug release in vivo. *Journal of Controlled Release*, Vol.147, No.3, (November 2010), pp. 350-358, ISSN 0168-3659

Peppas, N.A. & Sahlin, J.J. (1996). Hydrogels as mucoadhesive and bioadhesive materials: a review. *Biomaterials*, Vol. 17, No.16, 1996, pp. 1553-1561, ISSN 0142-9612

Porter, S.C. (1995). The Coating of Pharmaceutical Dosage Forms, In: *The Science and Practice of Pharmacy*, J.P. Remington, A.R. Gennaro, (Eds.), 1650-1659, Mack Pub., ISBN 0912734043, Easton, Pensylvania

Pushpadass, H.A., Marx, D.B. & Hanna, M. A. (2008). Effects of extrusion temperature and plasticizers on the physical and functional properties of starch films. *Starch/Stärke*, Vol.60, No.10, (October 2008), pp. 527-538, ISSN 1521-379X

Quaglia, F. (2008). Bioinspired tissue engineering: The great promise of protein delivery. *International Journal of Pharmaceutics*, Vol.364, No.2, (December 2008), pp. 281-297, ISSN 0378-5173

Quiao, M., Chen, D., Hao, T., Zhao, X., Hu, H. & Ma, X. (2007). Effect of bee venom peptide-copolymer interactions on thermosensitive hydrogel delivery systems. *International Journal of Pharmaceutics* Vol.345, No.1-2, (December 2007), pp. 116-124, ISSN 0378-5173

Qussi, B., Suess, W. G. (2006). The Influence of Different Plasticizers and Polymers on the Mechanical and Thermal Properties, Porosity and Drug Permeability of Free Shellac Films. *Drug Development & Industrial Pharmacy*, Vol.32, No.4, (January 2006), pp. 403 – 412, ISSN 1520-5762

Rao, P.R. & Divan, P.V. (1997) Permeability studies of cellulose acetate free films for transdermal use: Influence of plasticizers. *Pharmaceutica Acta Helvetiae*, Vol.72, No.1, (February 1997), pp. 47-51, ISSN 0031-6865

Rowe, R.C. (1977). The adhesion of film coatings to tablet surfaces - the effect of some direct compression excipients and lubricants. *Journal of Pharmacy and Pharmacology*, Vol.29, No.1, (September 1977), pp. 723–726, ISSN 2042-7158

Rowe, R.C. (1981). The adhesion of film coatings to tablet surfaces - a problem of stress distribution. *Journal of Pharmacy and Pharmacology*, Vol.33, No.1 (September 1981), pp. 610–612, ISSN 2042-7158.

Rowe, R.C. (1992). Defects in film-coated tablets: Etiology and solutions, In: *Advances in Pharmaceutical Sciences*, D. Ganderton, T. Jones (Eds.), 65–99, Academic Press, ISBN, New York, USA

Rungseevijitprapa, W., Brazeau, G.A., Simkins, J.W. & Bodmeier, R. (2008). Myotoxicity studies of O/W-in situ forming microparticle systems. *European Journal of Pharmaceutics and Biopharmaceutics*, Vol. 69, No.1 (May 2008) pp. 126 – 133, ISSN 0939-6411

Sato, Y. M., Wang, Y.M., Adachi, I. & Horikoshi, I. (1996). Pharmacokinetic study of taxol-loaded poly(lactic-co-glycolic acid) microspheres containing isopropyl myristate after targeted delivery to the lung in mice. *Biological & Pharmaceutical Bulletin*, Vol.19, No.12 (December 1996), ISSN 0918-6158

Sawicki, W. & Lunio, R. (2005). Compressibility of floating pellets with verapamil hydrochloride coated with dispersion Kollicoat SR 30 D. *European Journal of Pharmaceutics and Biopharmaceutics*, Vol.60, No.1 (May 2005), pp.153 – 158, ISSN 0939-6411

Shin, S.-Ch. & Choi, J.-S. (2005). Enhanced efficacy of triprolidine by transdermal application of the EVA matrix system in rabbits and rats. *European Journal of Pharmaceutics and Biopharmaceutics*, Vol.61, No. 1-2, (September 2005), pp.14-19, ISSN 0939-6411

Schaefer, M. J. & Singh, J. (2000). Effect of isopropyl myristic acid ester on the physical characteristics and in vitro release of etoposide from PLGA microspheres. *AAPS PharmSciTech*, Vol.1, No.4, Article 32 (November 2000), ISSN 1530-9932

Schaefer, M. J. & Singh, J. (2002). Effect of tricaprin on the physical characteristic and in vitro release of etoposide from PLGA microspheres. *Biomaterials*, Vol.23, No.16 (August 2002), pp. 3465 - 3471, ISSN 0142-9612

Schilling, S.U., Shah, N.H., Malick, A.W., Infeld, M.H. & McGinity, J.W. (2007). Citric acid as a solid-state plasticizer for Eudragit RS PO. *Journal of Pharmacy and Pharmacology*, Vol.69, No.11, (November 2007), pp. 1493–1500, ISSN 0022-3573

Siepmann, J., Lecomte, F. & Bodmeier, R. (1999). Diffusion-controlled drug delivery systems: calculation of the required composition to achieve desired release profiles. *Journal of Controlled Release*, Vol.60, No.3, (June 1999), pp. 379–389., ISSN 0168-3659

Skultety, P.F. & Sims, S.M. (1987). Evaluation of the loss of propylene glycol during aqueous film coating. *Drug Delivery and Industrial Pharmacy, Vol.*13, No.12, (January 1987), pp. 2209-2219, ISSN 1520-5762

Smart, J.D. (2005). The basics and underlying mechanisms of mucoadhesion. *Advanced Drug Delivery Reviews*. Vol.57, No.11, (November 2005), pp. 1556– 1568, ISSN 0169-409X

Snejdrova, E. & Dittrich, M. (2009). Adhesivity of branched plasticized oligoesters. *Ceska a Slovenska Farmacie*, Vol.58, No.5-6, (December 2009), pp. 212 - 215, ISSN 1210-7816

Snejdrova, E. & Dittrich, M. (2008). The reological properties of plasticized branched oligoesters. *Folia Pharmaceutica Universitatis Carolinae*, Vol.XXXVII, pp. 21-25, ISSN 1210-9495

Sogol, K. & Ismaeil, H. (2011). Mechanical influence of static versus dynamic loadings on parametrical analysis of plasticized ethyl cellulose films. *International Journal of Pharmaceutics*, Vol.408, No.1-2, (April 2011), pp. 1-8, ISSN 0378-5173

Solanki, H.K., Thakkar, J.H. & Jani, G.K. (2010). Recent advances in implantable drug delivery. *International Journal of Pharmaceutical Sciences Review and Research*, Vol.4, No.3, Article 028, pp. 168-177, ISSN 0976-044X

Solorio, L., Babin, B.M., Patel, R.b., Mach, J., Azar, N. & Exner, A.A. (2010). Noninvasive characterisation of in situ forming implants using diagnostic ultrasound. *Journal of Controlled Release*, Vol.143, No.2, (April 2010), pp. 183-190, ISSN 0168-3659

Somwanshi, S. B., Dolas, R.T., Nikam, V.K., Gaware, V.M., Kotade, K.B., Dhamak, K.B., Khadse, A.N. & Kashid, V.A. (2011). Effects of drug-polymer ratio and plasticizer concentration on the release of metoprolol polymeric microspheres. *International Journal of Pharmaceutical Research and Development*, Vol.3, No.3, (May 2011), pp. 139 - 146, ISSN 0974-9446

Steele, T. W. J., Huang, Ch.L., Widjaja, E., Boey, F.Y.C., Loo, J.S.C. & Venkatraman, S.S. (2011). The effect of polyetylene glycol structure on paclitaxel drug release and mechanical properties of PLGA thin films. *Acta Biomaterialia*, Vol.7, No.5, (May 2011), pp. 1973-1983, ISSN 1742 - 7061

Tang, Y. & Singh, J. (2009). Biodegradable and biocompatible thermosensitive polymer based injectable implant for controlled release of protein. *International Journal of Pharmaceutics*, Vol.365, No.1-2, (January 2009), pp. 34-43, ISSN 0378-5173

Udipi, K. & Cheng, P. (2004). Plasticized stent coatings. US Pat. Appl. No. 0215336 (October 28, 2004)

Warren, S.L., Dadey, E., Dunn, R.L., Downing, J.M. & Li, E.Q. (2009). Sustained delivery formulations of octreotide compounds. US Pat. Appl. No. 20090092650 (April 09, 2009)

Warren, S.L., Osborne, D.W. & Holl, R. (2008). Adhesive bioerodible ocular drug delivery system. US Pat. Appl. No. 20080268021 (October 30, 2008)

Winzenburg, G., Schmidt, C., Fuchs, S. & Kissel, T. (2004). Biodegradable polymers and their potential use in parenteral veterinary drug delivery systems. *Advanced Drug Delivery Reviews*, Vol.56, (February 2004), pp. 1453-1466, ISSN 0169-409X

Yuksel, N., Baykara, M., Shirinzade, H. & Suzen, S. (2011). Investigation of triacetin effect on indomethacin release from poly(methyl methacrylate) microspheres: Evaluation of interactions using FT-IR and NMR spectroscopies. *International Journal of Pharmaceutics,* Vol.404, No.1-2 (February 2011), pp. 102-109, ISSN 0378-5173

Zhu, Y., Shah, N.H., Malick, A.W., Infeld, M.H. & McGinity, J.W. (2002). Solid-state plasticization of an acrylic polymer with chlorpheniramine maleate and triethyl citrate. *International Journal of Pharmaceutics, Vol.* 241, No.2, (July 2002) pp. 301–310, ISSN 0378-5173

5

Plasticizers in Transdermal Drug Delivery Systems

Sevgi Güngör, M. Sedef Erdal and Yıldız Özsoy
Istanbul University Faculty of Pharmacy
Department of Pharmaceutical Technology, Beyazıt- Istanbul
Turkey

1. Introduction

Transdermal delivery is one of the non-invasive methods for drug administration. Patient compliance is improved and continuous, sustained release of drug is achieved by following the application of transdermal formulation on the skin (Guy 1996; Tanner & Marks 2008). Transdermal drug delivery systems, known as patches, are dosage forms designed to deliver a therapeutically effective amount of drug across a patient's skin in a predetermined time and controlled rate (Aulton 2007; Tiwary et.al., 2007; Vasil'ev et.al., 2001).

Transdermal drug delivery systems can be divided into three main groups : a) adhesive systems, in which the drug in adhesive, b) matrix type systems in which the drug in a matrix polymer and c) reservoir systems (Delgado-Charro & Guy 2001; Williams, 2003). Although there are differences in the design of transdermal therapeutic systems, several features are common to all systems including the release liner, the pressure sensitive adhesive, and the backing layer (Walters and Brain, 2007).

There are three critical considerations in the selection of a transdermal drug delivery system: adhesion to skin, compatibility with skin, and physical or chemical stability of total formulation and components (Walters and Brain, 2007). The adhesive nature of the patches is critical to the safety, efficacy, and quality of the product. Therefore the three important performance tests to monitor adhesive performance of patches are tack, shear strength and peel adhesion (Gutschke et al., 2010; Patel and Baria 2011; Ren et al., 2009). The choice and design of polymers, adhesives, penetration enhancers and plasticizers in transdermal patches are also critical because they have a strong effect on drug release, permeability, stability, elasticity, and wearing properties of transdermal drug delivery systems (Quan, 2011).

Plasticizers are low molecular weight resins or liquids, which cause a reduction in polymer-polymer chain secondary bonding, forming secondary bonds with the polymer chains instead (Gal and Nussinovitch, 2009; Rajan et al., 2010). The reasons for the use of plasticizers in transdermal drug delivery systems are the improvement of film forming properties and the appearance of the film, decreasing the glass transition temperature of the polymer, preventing film cracking, increasing film flexibility and obtaining desirable mechanical properties (Wypych, 2004). One of the many advantages of plasticizers used in transdermal formulations is the controlling of the release rate of therapeutic compound which can be done by the selection of the plasticizer type and the optimization of its concentration in the formulation. The commonly used plasticizers in transdermal patches include phthalate esters, phosphate esters, fatty acid esters and glycol derivatives (Bharkatiya et al, 2010; Wypych, 2004).

The objectives of this chapter are to summarize the compositions and types of the transdermal drug delivery systems; to emphasize the role and effectiveness of plasticizers in transdermal drug delivery systems and to cover the research studies and current developments related to the development of transdermal formulations.

2. Transdermal drug delivery systems

Transdermal drug delivery systems, also known as "patches", are dosage forms designed to deliver a therapeutically effective amount of drug across a patient's skin in a predetermined time at controlled rate and to maintain constant drug plasma concentration over a long period (Aulton 2007; Vasil'ev et.al., 2001). Transdermal patches have superiorities such as improved patient compliance and flexibility of dosage in which formulation can be removed immediately (Brown et.al. 2006; Guy 1996; Tanner & Marks 2008; Williams 2003). The transdermal systems of drugs including scopolamine, nitroglycerin, isosorbide dinitrate, clonidine, estradiol, fentanyl, nicotine, testosterone, norelgestromin+ethinyl estradiol, oxybutynin, selegeline, methylphenidate, buprenorphine, rivastigmine, rotigotine and granisetron have been approved (Guy 2010).

The major technical considerations by developing a transdermal formulation include (Meathrel, 2011; Quan, 2011):

- Size of the drug molecule and the required daily dose
- Drug compatibility with polymers, adhesives, plasticizers and other excipients used in the formulation
- Physical and chemical stability of the final formulation
- Size of the patch
- Balance between adhesion and easy patch removal depending on the duration of patch application.

2.1 Types of transdermal drug delivery systems

Although transdermal systems are classified in different types, transdermal patches can be divided into three main categories depending on the incorporation style of the drug in the system (Figure 1): (Delgado-Charro & Guy 2001; Padula et.al. 2007; Vasil'ev et.al., 2001; Williams, 2003).

- Reservoir Systems
- Matrix Systems
- Adhesive Systems

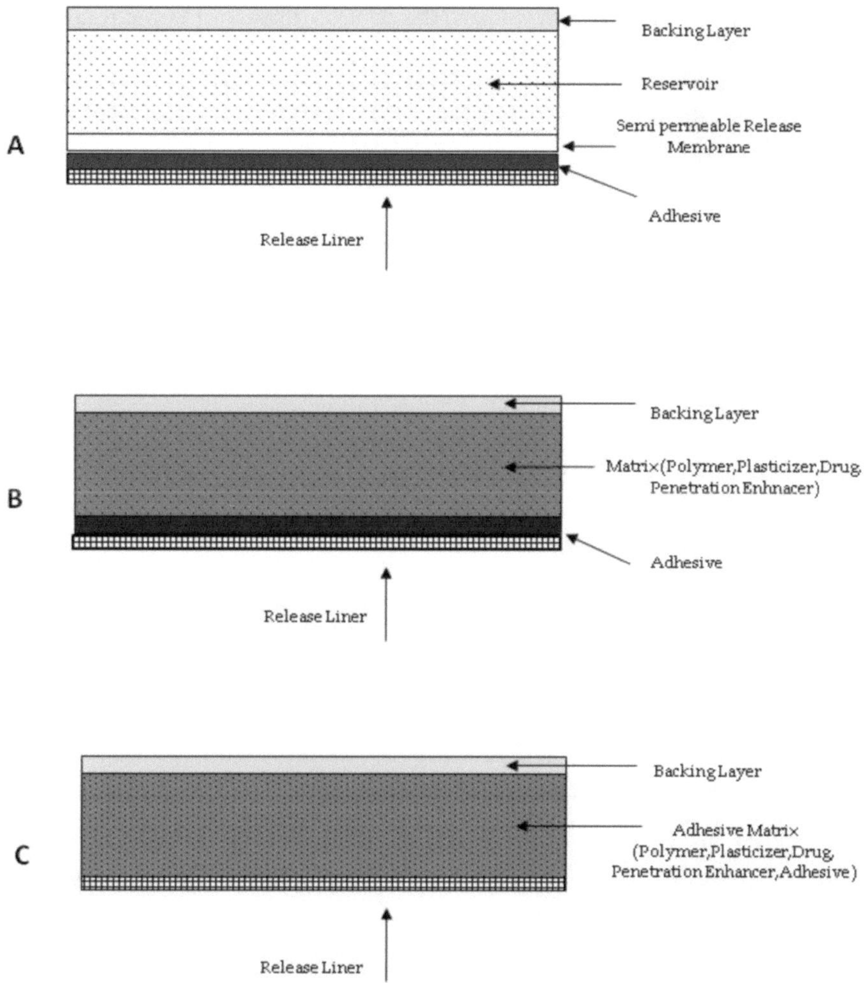

Fig. 1. Schematic representation of transdermal patch types: A. Reservoir, B. Matrix, C. Drug-in-Adhesive transdermal systems.

a. **Reservoir Systems:** In these systems, the drug is in a reservoir as liquid. Drug molecules are contained in the storage part, as a suspension in a viscous liquid or dissolved in a solvent. In the second type, there is a membrane made of a polymer with different structure, which separates the reservoir from the adhesive layer. In these systems, the membrane controls the release rate of the drug. The membrane can be porous or nonporous. The adhesive polymer on the exterior surface of the membrane enables the transdermal to adhere to skin. In these systems, drug release rate can be controlled by membrane thickness and adhesive layer (Delgado-Charro & Guy 2001; Padula et.al. 2007; Williams, 2003).

Transderm-Nitro (Nitroglycerin), Transderm-Scop (Scopolamine), Catapress-TTS (Clonidine), Estraderm (Estradiol) can be given as examples to the commercially available membrane diffusion controlled systems.

b. **Matrix Systems:** In this type of systems, the drug is dispersed homogeneously within a polymer matrix which has hydrophilic or lipophilic character. Outer side of the formulation is covered with a backing layer. In these systems, patch is held on the skin with a adhesive polymer as a strip. Matrix type formulations can also be prepared by dispersing the drug in an adhesive polymer that is sensitive to direct pressure and then covering this system with an impermeable backing layer. Since in matrix type formulations, release from semi-solid matrix of the drug is not controlled by any membrane, drug release from these systems is related to the surface area to which the patch is applied (Delgado-Charro & Guy 2001; Padula et.al. 2007; Williams, 2003).

Minitran (Nitroglycerin), Emsam (Selegeline), Exelon (Rivastigmine), Sancuso (Granisetron) and Oxytrol (Oxybutyne) can be given as examples to commercially available matrix diffusion controlled systems.

c. **Adhesive Systems:** In these systems, drug reservoir is prepared by dispersing the drug in an adhesive polymer. At the outmost, an impermeable backing layer takes place. Under the drug reservoir layer, there exists an adhesive membrane controlling the drug release rate. In this type of transdermal systems, drug release rate is controlled both by the matrix in which the drug is dispersed and also by a membrane. Although this type of systems can be designed with a single drug layer, they can be also designed as multi-layered (Delgado-Charro & Guy 2001; Padula et.al. 2007; Williams, 2003).

Nitrodur (Nitroglycerin), Daytarana (Methyl phenidate) and Duragesic (fentanyl) can be given as examples to commercially available adhesive systems.

2.2 Composition of transdermal drug delivery systems

Although transdermal systems can be design as different type systems mentioned above, following are the basic components which generally are used in the formulations of almost all type of transdermal patches (Williams, 2003).

- Drug
- Matrix
- Reservoir
- Semi-permeable (release) membrane
- Adhesive
- Backing layer
- Release liner
- Solvents, penetration enhancers
- Plasticizers

a. **Drug:** The drug, of which transdermal system will be designed, should possess some physicochemical characteristics. Drug should have relatively low molecular weight (<500 Dalton), medium level lipophilic character (log P 1-3.5) and water solubility (>100 mcg/ml). Also, the drug should be a potent compound, which is effective at low dose (<20 mg) (Guy 1996; Quan 2010).
b. **Matrix:** In the formulation of matrix type transdermal systems, the drug is dispersed or dissolved in a polymer matrix (Delgado-Charro & Guy 2001; Williams 2003). This matrix with polymer structure controls the release rate of the drug. Natural (e.g. pectin,

sodium alginate, chitosan), synthetic (Eudragit, polyvinyl pyrolidon, PVA) and semi-synthetic polymers (e.g. cellulose derivatives) are used as polymer (Amnuaikit et al., 2005; Güngör et al., 2008; Lin et al., 1991; Nicoli et al., 2006; Schroeder et al., 2007,a).

c. **Reservoir:** In this type of transdermal patches, a semi-permeable membrane controlling the drug release rate is used. The drug presents in a reservoir as liquid or solid (Delgado-Charro & Guy 2001; Williams 2003).

d. **Semipermeable (release) membrane:** It takes place in reservoir type transdermal systems and multi-layer adhesive systems. Ethylene-vinyl acetate copolymer, silicones, high-density polyethylene, polyester elastomers, cellulose nitrate and cellulose acetate are used as membrane. These membranes control the release rate of drugs (Williams 2003).

e. **Adhesive:** Adhesive should enable the transdermal system to easily adhere to the skin and should not be irritant/allergen for skin. Generally, pressure-sensitive adhesives are used in transdermal systems. Commonly used pressure-sensitive adhesives are collected under 3 classes as a) acylates, b) polyisobutilene adhesives and c) polysiloxan adhesives (Williams 2003).

f. **Backing layer:** It protects the system from external effects during administration and ensures integrity of the system in the storage period. For this purpose, the materials impermeable for drug molecule are used as backing layer. The backing layer must be inert and not compatible with the drug and other substances used in the formulation. Generally, ethylene vinyl acetate, polyethylene, polypropylene, polyvinylidene chloride and polyurethane are used as backing layer (Williams 2003).

g. **Release liner:** This is the part which protects the formulation from external environment and which is removed before the system is adhered to skin. Ethylene vinyl acetate, aluminum foil or paper can be used. Ideally, it should be easily peeled from the adhesive layer and should not damage the structure of adhesive layer. Also, silicone, fluorosilicone, perfluorocarbon polymers can be used (Williams 2003).

h. **Solvents, penetration enhancers:** Various solvents are used to solve or disperse the polymer and adhesive or drug used in preparation of the transdermal systems. Among those, chloroform, methanol, acetone, isopropanol and dichloromethane are used frequently. Also, various penetration enhancer substances are added to the formulations to increase permeation from skin of the drug. Terpenes, fatty acids, water, ethanol, glycols, surface-effective substances, azone, dimethyl sulfoxide are widely used in the transdermal formulations as permeation enhancer (Williams 2003).

i. **Plasticizers:** In transdermal systems, plasticizers are used to improve the brittleness of the polymer and to provide flexibility (Williams 2003).

3. Plasticizers

Plasticizers are generally non-volatile organic liquids or solids with low melting temperature and when added to polymers, they cause changes in definite physical and mechanical characteristics of the material (Bharkatiya et al, 2010; Felton, 2007; Gooch, 2010; Meier et al., 2004).

3.1 The role of plasticizers in pharmaceutical formulations

The main reasons of adding plasticizers to polymers, improving flexibility and processability are counted (Harper, 2006; Höfer & Hinrichs, 2010; Rahman & Brazel, 2004; Whelan, 1994). Upon addition of plasticizer, flexibilities of polymer macromolecules or macromolecular segments increase as a result of loosening of tightness of intermolecular forces (Bergo & Sobral, 2007; Höfer & Hinrichs, 2010).

The plasticizers with lower molecular weight have more molecules per unit weight compared to the plasticizers with higher molecular weight. These molecules can more easily penetrate between the polymer chains of the film forming agent and can interact with the specific functional groups of the polymer (Gal & Nussinovitch, 2009).

By adding plasticizer to a polymeric material, elongation at break, toughness and flexibility are expected to increase, on the other hand tensile stress, hardness, electrostatic chargeability, Youngs modulus and glass transition temperature are expected to decrease (Gal & Nussinovitch, 2009; Harper, 2006; Rahman & Brazel, 2004).

Plasticizers with low molecular weight, act by reducing the secondary bonds (e.g. hydrogen bond) of the polymer chains and themselves forming secondary bonds (Gal & Nussinovitch, 2009). While low molecular weight improves miscibility with the polymer, the second factor increasing the compatibility is the realization of strong mutual hydrogen bonding (Harper, 2006). Thus, weakening of interaction of the polymer chains decrease tensile strength and glass transition temperature and so the flexibility of polymer films increases (Felton, 2007; Rahman & Brazel, 2004).

3.2 Classification of plasticizers

Several substances, including water, can be used to plasticize the polymer. It is reported that, phthalate, sebacate and citrate esters are among the most commonly used plasticizers (Felton, 2007). Compatibility, general structure (being a monomeric or polymeric), functions and chemical structure are taken into account in classifying the plasticizer substances (Gooch, 2010).

Most used group of the plasticizer substances is the phthalic acid esters which have firstly put into use in 1920. Dioctyl-phthalate is the most commonly used phthalic acid ester and it constitutes 50% of the world's plasticizer consumption (Höfer & Hinrichs, 2010).

Aliphatic ester plasticizers are derived from esterification of adipic, sebacic and azelaic acids with linear or branched monofunctional alcohols of short or medium length of chain (e.g. dioctyladipate and dibutylsebacate). Adipate, azelate and sebacate plasticizers are distinguished from other groups by their low viscosity. They give flexibility to the polymers they are used together at low temperatures (Harper, 2006; Höfer & Hinrichs, 2010; Rahman & Brazel, 2004).

Phosphate esters and various glycol derivatives such as propylene glycol and polyethylene glycol are also employed to plastify the polymeric films (Felton, 2007; Harper, 2006; Meier et al., 2004; Rahman & Brazel, 2004).

It has been reported that, surfactants, preservatives and other compounds also function as plasticizer agent together with cellulosic and acrylic polymers (Felton, 2007).

3.3 Properties of plasticizers

A plasticizer is firstly expected to be compatible with the polymer substance. This means that, it can fully mix with the polymer and can remain permanently in the polymer.

Tendency to migration, exudation, evaporation or volatilization of the plasticizers employed in a polymeric system must be low (Felton, 2007; Harper, 2006).

Other properties expected from an ideal plasticizer are its workability, its ability to provide desired thermal-electrical and mechanical characteristics to the end product, its durability at high and low temperature values, its being effective over a wide temperature range and not being affected by ultraviolet radiation, its cost being low and its conformance to the health and safety arrangements (Rahman & Brazel, 2004).

3.4 Effectiveness of plasticizers

While some of approximately 600 commercial plasticizers are very effective in softening the polymers, the others do not exhibit efficiency in this area and are used for different purposes (Harper, 2006). To exhibit efficiency, the plasticizer should be able to transit from solvent phase to polymer phase and then it can diffuse into polymer and disrupt the intermolecular interactions (Felton, 2007).

Three factors determine the effectiveness of a plasticizer to be used with polymers (Harper, 2006):

1. A flexible plasticizer molecule with long $(CH_2)_n$ chains is more effective in increasing the polymer flexibility.
2. Low polarity and hydrogen bonding cause decrease in the interaction between the polymer and the plasticizer (borderline compatibility).
3. The plasticizers with low molecular weight are more active and they increase the flexibility more.

The forces affecting the polymer-plasticizer mixtures are identified as hydrogen bonds, dipole-dipole interactions and dispersion forces. The methods used in measuring the extent of polymer plasticizer interaction can be listed as follows (Felton, 2007):

- Torsion braid pendulum
- Vapor pressure depression
- Osmotic pressure
- Swelling tests
- Gas-liquid chromatography
- Viscometry
- Melting point depression
- Nuclear magnetic resonance (NMR) Spectroscopy
- Fourier Transform Infrared (FTIR) Spectrometry

Plasticizers are generally compared with a plasticizer with well known characteristics such as dioctylphthalate. A characteristic such as modulus or hardness is chosen and a value for this characteristic is determined. The ratio of plasticizer concentrations (test/dioctylphthalate) required to achieve this value is defined as the effectiveness of the plasticizer (Whelan, 1994). The effectiveness can also be measured by graphing the modulus versus plasticizer concentration and the graphs of various plasticizers can be compared (Harper, 2006). Most commonly used methods in measuring the effectiveness of the plasticizer are DSC analyses and the decrease in glass transition temperature when plasticizer is added to polymer (Felton, 2007; Zhu et al., 2002).

In pharmaceutical formulations, effectiveness of a plasticizer agent substantially depends on its amount added to the formulation and on the polymer-plasticizer interaction. When an aqueous dispersion is in question, the proportion and amount of partition was found to be dependent on the solubility of the plasticizer in water and its affinity to the polymer phase. When water-insoluble plasticizers will be dispersed in an aqueous medium, they should firstly be emulsified and then added to the polymer (Felton, 2007).

4. Plasticizers in transdermal drug delivery systems

Many of the polymers used in pharmaceutical formulations are brittle and require the addition of a plasticizer into the formulation. Plasticizers are added to pharmaceutical polymers intending to ease the thermal workability, modifying the drug release from polymeric systems and improving the mechanical properties and surface properties of the dosage form (Felton, 2007; Lin et al., 2000; Wang et al., 1997; Wu & McGinity, 1999; Zhu et al., 2002).

The plasticizers used in pharmaceutical formulations present a) in coating material of solid dosage forms, and b) in transdermal therapeutic systems. The list of frequently used plasticizers in pharmaceutical formulations is given below (Table 1) (Wypch, 2004).

Group	**Hydrophilic/Lipophilic**	**Plasticizer**
Glycerol and esters	Hydrophilic	Glycerine, Glycerine triacetate, Glyceryltributyrate
Glycol derivatives	Hydrophilic	Propylene glycol, Poliethylene glycol
Phthalic acid esters	Lipophilic	Dibutyl phthalate, Diethyl phthalate
Sebacic acid esters	Lipophilic	Dibutyl sebacate, Diethyl sebacate
Oleic acid esters	Hydrophilic	Oleil oleate
Sugar alcohols	Hydrophilic	Sorbitol
Citric acid esters	Hydrophilic	Triethyl citrate, Tributhyl citrate
Tartaric acid esters	Lipophilic	Diethyl tartarate

Table 1. Plasticizers Used in the Pharmaceutical Formulations (Wypch, 2004).

It is observed that the plasticizers added to transdermal therapeutic systems are mostly used in the proportions between 5-20%. The chemical formulas of 6 plasticizers frequently used in transdermal drug delivery studies are given in Table 2.

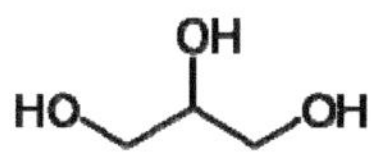 Glycerine	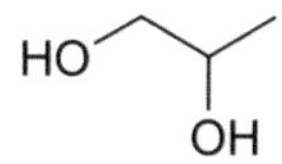 Propylene Glycol
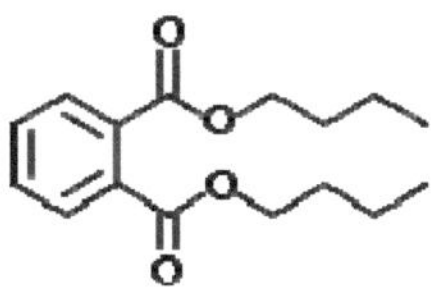 Dibutyl Phtalate	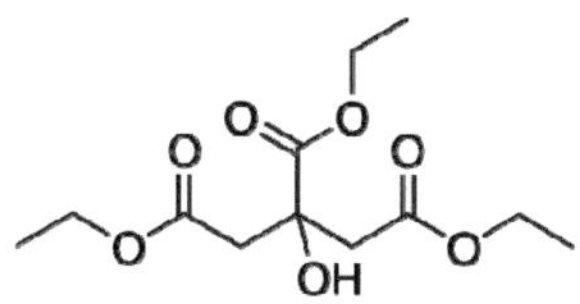 Triethyl citrate
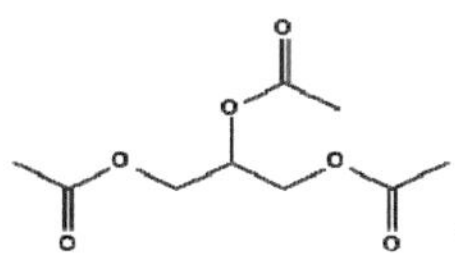 Triacetin	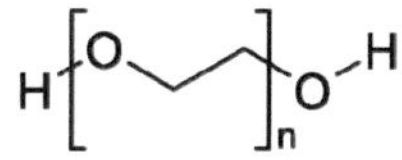 Polyethylene Glycol

Table 2. The chemical formulas of the frequently used plasticizers in transdermal formulations.

Among the plasticizers commonly used in the formulation of transdermal films, there are the phthalate and citrate esters and glycol derivatives (Gal & Nussinovitch, 2009). The plasticizers used in the studies conducted through last 20 years, their proportions and the polymers they are used with are given as a table (Table 3).

Following are the reasons which can be counted among those for adding plasticizers to the polymer films to be used in transdermal drug delivery systems:

- Reducing the brittleness
- Improving flow
- Ensuring flexibility
- Enhancing the resistance and tear strength of the polymer film (Bergo & Sobral, 2007; Felton, 2007; Rao & Diwan, 1997).

Plasticizer	% w/w	Polymer	Type of Transdermal Formulation	Reference
Triacetin	1.43-5.48	Eudragit E 100	Drug free film	Lin et al., 1991
Sorbitol Sucrose	20	Polyvinyl Alcohol: Chitosan	Drug free film	Arvanitoyannis et al., 1997
Dibutyl phthalate Propylene glycol Polyethylene glycol 600	40	Cellulose acetate	Drug free film	Rao & Diwan, 1997
Polyethylene glycol 3350	5	Hydroxypropylcellulose Hydroxypropylcellulose: Eudragit E 100 Hydroxypropylcellulose: Carbopol 971P Hydroxypropylcellulose: Polycarbophil	Matrix	Repka & McGinity, 2001
Polyethylene glycol 600	10-50	Cellulose Acetate	Membrane	Wang et al., 2002
Triethyl citrate Dibutyl phthalate	10	Eudragit E 100	Matrix	Gondaliya & Pundarikakshudu, 2003
Glycerine	4	Polyvinyl Alcohol 72000	Matrix	Padula et al., 2003
Polyethylene glycol 400	40	Carboxymethyl Guar	Matrix	Murthy et al., 2004
Dibutyl phthalate	30	Ethyl cellulose Polyvinylpyrrolidone	Matrix	Amnuaikit et al., 2005
Dibutyl phthalate	20	Ethyl cellulose: Polyvinylpyrrolidone Eudragit: Polyvinylpyrrolidone	Matrix	Mukherjee et al., 2005
Sorbitol Solution (70%)	2	Polyvinyl Alcohol	Matrix	Nicoli et al., 2005
Sorbitol Solution (70%)	4	Polyvinyl Alcohol 83400	Matrix	Femenia-Font et al., 2006
Sorbitol Solution (70%)	4	Polyvinyl Alcohol	Matrix	Nicoli et al., 2006
Dibutyl phthalate	30	Ethylcellulose: Polyvinylpyrrolidone	Matrix	Dey et al., 2007
Dibutyl phthalate Propylene Glycol	20	Polyvinyl Alcohol Xanthan Gum	Matrix	Kumar et al., 2007
Triethyl citrate	6	Hydroxypropylcellulose Eudragit RL PO Silicon Gum Acrylate copolymer	Film forming polymeric solution	Schroeder et al., 2007,a

Plasticizer	% w/w	Polymer	Type of Transdermal Formulation	Reference
Triethyl citrate Triacetin Dibutyl phthalate	1-6 2.1 4	Eudragit RL PO, E100, S100 and NE 40D Polyvinylpyrrolidone Hydroxypropylcellulose Acrylate copolymer Acrylate/Octylacrylamide copolymer Silicon Gum Polyvinyl Alcohol Polyisobutylene	Film forming polymeric solution	Schroeder et al., 2007, b
Tributyl citrate Triacetin	25-125	Eudragit NE40D	Matrix	Cilurzo et al., 2008
Propylene glycol	10	Pectin	Matrix	Güngör et al., 2008
Triacetin	20	Eudragit E 100: Eudragit NE40D	Matrix	Inal et al., 2008
Dibutyl phthalate Triethyl citrate	30	Hydroxypropyl methyl cellulose:Ethyl cellulose	Matrix	Limpongsa & Umprayn, 2008
Glycerin Polyethylene glycol 400 Propylene Glycol	20, 40	Polyvinyl Alcohol Polyvinylpyrrolidone	Matrix	Barhate et al., 2009
Triacetin	10-45	Eudragit E 100	Matrix	Elgindy & Samy, 2009
Glycerin Polyethylene glycol 200 Polyethylene glycol 400	10	Polyvinyl Alcohol : Polyvinylpyrrolidone	Matrix	Gal & Nussinovitch, 2009
Propylene Glycol Dibutyl phthalate	30 30	Polyvinyl Alcohol: Polyvinylpyrrolidone Ethyl cellulose: Polyvinylpyrrolidone	Matrix	Jadhav et al., 2009
Propylene Glycol Polyethylene glycol 400	5, 10	Eudragit L100 Eudragit L100-55 Eudragit S100	Matrix	Marzouk et al., 2009
Propylene Glycol	20, 30,40	Hydroxypropyl methyl cellulose Ethyl cellulose Carboxy methyl cellulose	Matrix	Pandit et al., 2009
Dibutyl phthalate	30	Ethyl cellulose: Polyvinylpyrrolidone Ethyl cellulose: Hydroxypropyl methyl cellulose	Matrix	Bagchi & Kumar Dey, 2010

Plasticizer	% w/w	Polymer	Type of Transdermal Formulation	Reference
Dibutyl phthalate Propylene Glycol Polyethylene glycol 400	40	Cellulose acetate Hydroxypropyl methyl cellulose Polyvinylpyrrolidone Polyethylene glycol 4000 Eudragit RL 100-RS 100	Matrix	Bharkatiya et al., 2010
Propylene Glycol	15	Hydroxypropyl methyl cellulose E15: Eudragit RS 100 Hydroxypropyl methyl cellulose E15: Eudragit RL 100	Matrix	Karunakar et al., 2010
Propylene Glycol	20	Hydroxypropyl methyl cellulose Eudragit RL 100	Bilayered Matrix	Madishetti et al., 2010
Propylene Glycol	20	Eudragit RL 100 Hydroxypropyl methyl cellulose	Matrix	Mamatha et al., 2010
Polyethylene glycol 400 Dibutyl phthalate Glycerin	15, 30,40	Gum Copal	Matrix	Mundada & Avari, 2010
Sorbitol Glycerin	4	Polyvinyl Alcohol 29, 83,115	Matrix	Padula et al., 2010
Dibutyl phthalate Dibutilsebacate	5, 10	Eudragit E100 Polyvinylpyrrolidone	Matrix	Rajabalaya et al., 2010
Dibutyl phthalate	5-25	Eudragit RS 100, Eudragit RL 100, Polyvinylpyrrolidone	Matrix	Rajan et al., 2010
Triethyl citrate	5	Hydroxypropyl methyl cellulose, Eudragit RL 100, Chitosan	Matrix	Shinde et al., 2010
Polyethylene glycol 400	5, 10	Eudragit RL 100 Eudragit RS 100	Matrix	Amgoakar et al., 2011
Glycerin	10	Polyvinyl Alcohol Polyvinylpyrrolidone Trimethoxysilane	Matrix	Guo et al., 2011
Glycerin	10, 20, 30	Hydroxypropyl methyl cellulose Polyvinylpyrrolidone Eudragit RS100	Matrix	Irfani et al., 2011
Glycerin	5	Polyvinyl Alcohol:Eudragit L30D55	Matrix	Nesseem et al., 2011
Dibutyl phthalate	10	Ethyl cellulose Hydroxypropyl methyl cellulose	Matrix	Parthasaraty et al., 2011

Table 3. The plasticizers used in the transdermal studies conducted through last 20 years, their proportions and the polymers they are used with.

The selection of plasticizers depends on the characteristics of polymer used to prepare transdermal formulation. When the composition of transdermal film/patch formulations in the patents was looked over, it was seen that polyvalent alcohols *e.g.* glycerin and 1,2 propandiol (propylene glycol) are generally used as plasticizers to softening the polymers in the formulation (Deurer et.al. 1999; Herrmann & Hille, 1999; Selzer 2001; Selzer, 2004). Higher alcohols such dodecanol, or mineral oil, silicone oil, isopropyl myristate; isopropyl palmitate; polyethylene glycol 400; diethyl sebacate and/or dibutyl sebacate; hydrocarbons, alcohols, carboxylic acids and derivatives thereof were also added into transdermal formulations as plasticizer (Petereit et.al. 2005; Salman & Teutsch 2011; Selzer, 2001).

Without a plasticizer, a very hard but brittle film is obtained. This means that, external forces such as bending, stretching and stripping from surface will cause tearing of the film without too much effort. However, when transdermal patches are in question, rather than reduction in the hardness of the patch, its endurance when positioned or repositioned on the skin is important (Gal & Nussinovitch, 2009).

In studying the mechanical properties of transdermal patches or films, tensile testing is the primarily interested subject. Tensile tests enable to study the mechanical properties of the formulation such as stress strain curves and stress at failure. These properties provide information about the resistance to damage during storage and usage. The effect of the type and proportion of the plasticizer in a formulation on the mechanical properties can also be understood by this way (Gal & Nussinovitch, 2009; Rajabalaya et al., 2010).

The tensile strength of the transdermal films varies with the type of the polymer and plasticizer used. Generally a soft and weak polymer is identified with low tensile strength and low elongation values, a hard and brittle polymer is identified with moderate tensile strength and low elongation values and a soft and tough polymer is identified with high tensile strength and high elongation values (Bharkatiya et al., 2010).

Barhate et al. have prepared matrix type transdermal patches using polyvinyl alcohol and polyvinyl pyrolidone as polymer and using glycerin, polyethylene glycol 400 and propylene glycol in proportions 20% and 40% as plasticizer and studied carvedilol permeation from these patches. Plasticizers used have ethylene oxide groups and display their effects thanks to the hydrogen bonds they form with polymer molecules. This interaction gives flexibility to the polymer. Tensile strength measures the ability to patch to withstand rupture. In the formulations prepared, highest tensile strength has obtained when glycerine was used as plasticizer. On the other hand, it was determined that in vitro permeation of carvedilol increased when polyethylene glycol 400 was used (Barhate et al., 2009).

Drug free polymeric patches have been prepared using various polymers (Eudragit, hydroxypropylmethyl cellulose, cellulose acetate, polyvinyl pyrolidone and polyethylene glycol 4000) and effect of various plasticizers on mechanical and physicochemical properties of the patches have been investigated. Polyethylene glycol 400, dibutyl phthalate and propylene glycol were used as plasticizer in proportion of 40% (w/w) of the weight of the dry polymer. Tensile strength and folding endurance properties of the patches prepared with dibutyl phthalate have been found higher compared to those prepared with propylene glycol and polyethylene glycol 400 (Bharkatiya et al., 2010).

In a study conducted with cellulose acetate transdermal films, it has been determined that when dibutyl phthalate and polyethylene glycol 600 have been used as plasticizer, the transparency of the films were not differing from the films not containing plasticizer, on the other hand, when propylene glycol has been used, it created a light opaqueness. Besides, the flexibility of the films plasticized with 40% plasticizer has been determined to be much better than unplasticized films and they could be removed without rupture from the surface they were adhered to (Rao & Diwan, 1997).

In a study where polyvinyl alcohol and polyvinyl alcohol-Xanthan gum mixture has been used as polymer, propylene glycol and dibutyl phthalate have been chosen as plasticizer. It has been observed that, addition of xanthan gum and dibutyl phthalate to the films prepared with only polyvinyl alcohol decreases the tensile strength and increases the percentage elongation. On the other hand, in polyvinyl alcohol films prepared with propylene glycol, tensile strength has been found higher. Besides, in vitro release of the terbutaline sulphate, which is an active substance, has been found higher in the propylene glycol containing films (Kumar et al., 2007).

Eudragit E 100, is a good polymer candidate in preparing transparent and self-adhesive transdermal films. However, its mechanical properties should be enhanced by adding a plasticizer. In a study evaluating drug free Eudragit E 100 films, it has been observed that elongation value of the films has increased depending on the increase of concentration of plasticizer. Triacetin used as plasticizer in combination with a cohesion promoter, succinic acid. It is thought that, the plasticizer ensures this effect with lubrication of the polymer chains. Differential Scanning Calorimetry (DSC) analyses have shown that, the crystallinity was decreased in plasticized Eudragit E 100 films when compared with those not plasticized. Increase in the mobility of the polymer chains and corresponding decrease in the crystallized area existing within the polymer are expected to enhance the active substance permeation. It has been concluded that, best cohesion promoter-plasticizer combination for Eudragit E 100 films was 7% succinic acid and 25% or 45% triacetine (Elgindy & Samy, 2009).

Although triacetin is considered as a good plasticizer for Eudragit E transdermal films, it has been determined that, addition of a secondary plasticizer such as polyethylene glycol 200, propylene glycol, diethyl phthalate or oleic acid to the system positively affects the transparency, flexibility and adhesive properties of the film (Lin et al., 1991).

Addition of plasticizer to the transdermal therapeutic systems may exhibit a facilitating effect in adhesion of the film to the other surfaces or membranes, by affecting the adhesiveness of the system (Gal & Nussinovitch, 2009; Rao & Diwan, 1997). Again in transdermal systems, humidity content and water absorption capacity of the system are measured and effect of the plasticizers on these values is researched (Rajabalaya et al., 2010). Water vapor transmission rate is closely related with the permeability characteristics of the transdermal films and can change according to the plasticizer and polymer type used (Bharkatiya et al, 2010).

It has been reported that, the plasticizers such as glycerine, sorbitol and polyethylene glycol can change release rate of the therapeutic components contained in the formula of transdermal drug delivery systems. Release rate of the drug can be adjusted by changing the type and concentration of the plasticizer (Lin et al., 2000; Wypch, 2004). Also it has been found that, the effect of the plasticizer on drug transport is related to the physicochemical properties of the permeant, in particular to its solubility in the plasticizer (Padula et al., 2010).

In transition studies conducted with diltiazem hydrochloride and indomethacine, diffusion of the drug has been found as the films plasticized with polyethylene glycol 600 > dibutyl phthalate > propylene glycol, in order. It has been concluded that, permeation of the drugs and mechanical properties of the film were affected by the choice of suitable plasticizer and its concentration (Rao and Diwan, 1997).

In their study, Amgokar and coworkers have prepared transdermal films of budesonide. In the films, Eudragit RL 100: Eudragit RS and ethylcellulose-polyvinylpyrolidone (in proportions of 7:3 and 7:2, respectively) have been used as polymers and polyethylene glycol 400 has taken place as plasticizer. While drug release has found proportional to the polymer concentration, increase in the plasticizer amount has caused an increase in the weight of the film. It has also been observed that, increase in the plasticizer amount has also increased the humidity absorption of the transdermal films. It has been reported that, permeation of budesonide was the highest when polyethylene glycol 400 has been used in proportion of 10% (Amgaokar et al., 2011).

Irfani and coworkers have used different combinations of hydoxypropyl methylcellulose, Eudragit RS 100 and polyvinylpyrolidone, when preparing transdermal films of the active substance, valsartan. In the formulations, different proportions (10%, 20% and 30%) of glycerin have been tried as plasticizer. It has been found that, increasing plasticizer concentration was increasing the diffusion rate of the active substance. Besides, when also the polymer combination of Eudragit RS and hydroxypropyl methylcellulose has been used with 30% glycerine, an increase in the diffusion rate of valsartan has been determined (Irfani et al., 2011).

In a study, effects of two plasticizers, dibutyl phthalate and dibutyl sebacate, on the mechanical properties of the transdermal films prepared with Eudragit 100-polyvinylpyrolidone polymer mixture have been researched. It was shown that tensile strength was gradually decreased as the plasticizer concentration in the patch increased. It can be concluded from this result that plasticizer molecules disrupt the inter-chain cohesive forces of the polymer. Dibutyl phthalate and dibutyl sebacate have affected the mechanical properties of the transdermal system similarly. The finding that dibutyl sebacate is a suitable plasticizer for a more rapid release has been found in conformance with the finding of Siepmann and colleagues stating that dibutyl sebacate ensures faster release, whereas plasticizers containing phthalate group should be preferred when extended effect is required (Rajabalaya et al., 2010; Siepmann et al., 1999).

In the study where Gum copal has been used as polymer, hydrophobic dibutyl phthalate and hydrophilic glycerin and polyethylene glycol 400 have been preferred as plasticizer. It has been observed that, the films prepared by using dibutyl phthalate were more homogeneous and clear and also, their tensile strength and % elongation values have been found higher. In the study where verapamil hydrochloride has taken place as active substance, the release has realized longer and more controlled than the films containing 30% dibutyl phthalate (Mundada and Avari, 2010).

The characterization of the cellulose membranes where polyethylene glycol 600 has been used as plasticizer has been made and it has been determined that, besides the plasticizer concentration, preparation temperature was also effective on the membrane properties. It has been determined that, the membranes prepared at 40°C were more homogeneous and the diffusion of the active substance scopolamine, has realized through 3 days, controlled and constant, from the membranes containing 10% or 20% polyethylene glycol 600. It has been reported that, in order to improve the mechanical properties of the cellulose acetate membranes and to enable the linear release of the active substance, polyethylene glycol concentration should be optimized (Wang et al., 2002)

5. Conclusion

There are several considerations in the optimization of a transdermal drug delivery system. The choice and design of polymers, adhesives, penetration enhancers and plasticizers in transdermal systems are crucial for drug release characteristics as well as mechanical properties of the formulation. Beside the other components of transdermal patches, plasticizers also significantly change the viscoelastic properties of the polymers. The reasons for the use of plasticizers in transdermal drug delivery systems are the improvement of film forming properties and the appearance of the film, preventing film cracking, increasing film flexibility and obtaining desirable mechanical properties. Therefore, the selection of the plasticizer type and the optimization of its concentration in the formulation should be carefully considered.

6. References

Amgoakar, Y.M.; Chikhale, R.V.; Lade, U.B.; Biyani, D.M. & Umekar, M.J. (2011). Design, Formulation and Evaluation of Transdermal Drug Delivery System of Budesonide. *Digest Journal of Nanomaterials and Biostructures*, Vol.6, No.2, (April-June 2011), pp.475-497, ISSN 1842-3582

Amnuaikit, C.; Ikeuchi, I.; Ogawara, K.; Higaki, K. & Kimura, T. (2005). Skin Permeation of Propranolol From Polymeric Film Containing Terpene Enhancers for Transdermal Use. *International Journal of Pharmaceutics*, Vol. 289, pp. 167-178, ISSN 0378-5173

Aqil, M.; Sultana, Y.; Ali, A.; Dubey, K.; Najmi, A.K. & Pillal, K.K. (2004). Transdermal Drug Delivery Systems of a Beta Blocker: Design, In Vitro, and ın Vivo Characterization. *Drug Delivery*, Vol.11, No.1, pp. 27-31, ISSN 1071-7544

Arvanitoyannis, I.; Kolokuris, I.; Nakayama, A.; Yamamoto, N. & Aiba, S. (1997). Physicochemical Studies of Chitosan-Polyvinylalcohol Blends Plasticized With Sorbitol and Sucrose. *Carbohydrate Polymers*, Vol.34, pp. 9-19, ISSN 0144-8617

Aulton, M.E. (2007). *Aulton's Pharmaceutics: The Design and Manufacture of Medicines. (3rd ed.)*, K. Taylor (Ed), 565-597, Elsevier, ISBN : 978-0-443-10107-6, London, UK

Bagchi, A. & Dey, B.K. (2010). Formulation, In Vitro Evaluations and Skin Irritation Study of Losartan Potassium Transdermal Patches. *Iranian Journal of Pharmaceutical Sciences*, Vol.6, No.3, pp. 163-170, ISSN 1735-2444

Barhate, S.D. (2009). Formulation and Evaluation of Transdermal Drug Delivery System of Carvedilol. *Journal of Pharmacy Research*, Vol.2, No.4, (April 2009), pp. 663-665, ISSN 0974-6943

Bharkatiya, M.; Nema, R.K. & Bhatnagar, M. (2010). Designing and Characterization of Drug Free Patches for Transdermal Application. *International Journal of Pharmaceutical Sciences and Drug Research*, Vol.2, No.1, pp. 35-39, ISSN 0975-248X

Bergo, P.V.A.; Sobral, P.J.A. (2007). Effects of Plasticizer on Physical Properties of Pig Skin Gelatin Films. *Food Hydrocolloids*, Vol.21, No.8, pp.1285-1289, ISSN 0268-005X

Brown, M.B., Martin, G.P., Jones, S.A., Akomeah, F.K. (2006). Dermal and Transdermal Drug Delivery Systems: Current and Future Prospects. *Drug Delivery*, Vol.13, No.3, (May 2006), pp.175-187, ISSN 1071-7544

Cilurzo, F.; Minghetti, P.; Pagani, S.; Casiraghi, A. & Montanari, L. (2008). Design and Characterization of an Adhesive Matrix Based on a Poly (Ethyl Acrylate, Methyl Metacrylate). *American Association of Pharmaceutical Scientists Pharmaceutical Science and Technology*, Vol.9, No.3, (September 2008), pp.748-754, ISSN 1530-9932

Delgado-Charro, M.B. & Guy, R.H. (2001). Transdermal Drug Delivery, In: *Drug Delivery and Targeting for Pharmacists and Pharmaceutical Scientists*. A.M. Hillery, A.W. Lloyd, J. Swarbrick (Ed.), 189-214, Taylor&Francis, ISBN 0-4152-7198-3, London, UK.

Deurer, L.; Otto, K., Hille, T. (1999). Systems for the Controlled Release of Pilocarpine. US005869086A

Dey, B.K.; Nath, L.K.; Mohanti, B. & Bhowmik, B.B. (2007). Development and Evaluation of Propranolol Hydrochloride Transdermal Patches by Using Hydrophilic and Hydrophobic Polymer. *Indian Journal of Pharmaceutical Education & Research*, Vol.41, No.4, (October-December 2007), pp.388-393, ISSN 0019-5464

Elgindy, N. & Samy, W. (2009). Evaluation of the Mechanical Properties and Drug Release of Cross Linked Eudragit Films Containing Metronidazole. *International Journal of Pharmaceutics*, Vol. 376, (March 2009), pp. 1-6, ISSN 0378-5173

Felton, L.A. (2007). Film Coating of Oral Solid Dosage Forms, In: *Encyclopedia of Pharmaceutical Technology*, J. Swarbrick (Ed.), 1729-1747, Informa Healthcare, ISBN 0-8493-9396-5, NY, USA

Femenia-Font, A.; Padula, C.; Marra, F.; Balaguer-Fernandez, C.; Merino, V.; Lopez-Castellano, A.; Nicoli, S. & Santi, P. (2006). Bioadhesive Monolayer Film for the In Vitro Transdermal Delivery of Sumatriptan. *Journal of Pharmaceutical Sciences*, Vol.95, No.7, (July 2006), pp. 1561-1569, ISSN 1520-6017

Gal, A. & Nussinovitch, A. (2009). Plasticizers in the Manufacture of Novel Skin Bioadhesive Patches. *International Journal of Pharmaceutics*, Vol. 370, No.1-2, (March 2009), pp. 103-109, ISSN 0378-5173

Gooch, J.W. (2010). *Encyclopedic Dictionary of Polymers*, Springer Reference, ISBN 978-1-4419-6246-1, NY, USA

Gondoliya, D. & Pundarikakshudu, K. (2003). Studies in Formulation and Pharmacotechnical Evaluation of Controlled Release Transdermal System of Bupropion. *American Association of Pharmaceutical Scientists Pharmaceutical Science and Technology*, Vol.4, No.1, (January 2003), pp. 1-9, ISSN 1530-9932

Guo, R.; Du, X.; Zhang, R.; Deng, L.; Dong, A. & Zhang, J. (2011). Bioadhesive Film Formed From a Novel Organic-Inorganic Hybrid Gel for Transdermal Drug Delivery System. *European Journal of Pharmaceutics and Biopharmaceutics*, Vol.79, No.3, (November 2011), pp. 547-583, ISSN 0939-6411

Gutschke, E; Bracht, S; Nagel, S. & Weitschies, W. (2010). Adhesion Testing of Transdermal Matrix Patches With a Probe Tack Test in Vitro and in Vivo Evaluation. *Eur J Pharm Biopharm.*, Vol. 75, No.3, (Aug 2010), pp. 399-404, ISSN 0939-6411

Guy, R.H. (1996). Current Status and Future Prospects of Transdermal Drug Delivery. *Pharmaceutical Research*, Vol.13, No.12, (Aug 1996), pp. 1765-1769 ISSN 0724-8741

Guy, RH. (2010). Transdermal Drug Delivery. In: *Drug Delivery, Handbook of Experimental Pharmacology*, M. Schäfer- Korting (Ed.), Springer Verlag, ISBN: 978-3-642-00477-3-11-13, Berlin, Germany.

Güngör, S.; Bektas, A.; Alp, F.I.; Uydes-Dogan, B.S.; Özdemir, O.; Araman, A. & Özsoy, Y. (2008). Matrix Type Transdermal Patches of Verapamil Hydrochloride: In vitro Permeation Studies Through Excised Rat Skin and Pharmacodynamic Evaluation in Rats. *Pharmaceutical Development and Technology*, Vol.13, No.4, (January 2008), pp. 283-289, ISSN 1083-7450

Harper, C.A. (2006). *Handbook of Plastic Technologies, The Complete Guide to Properties and Performance*, Mc Graw-Hill Handbooks, ISBN 0-07-146068-3, NY, USA

Herrmann, F. & Hille T. (1999). Transdermal Therapeutic System with Pentylene Terazole as Active Substance. US 005965155.

Höfer, R. & Hinrichs, K. (2010). Additives for the Manufacture and Processing of Polymers, In: *Polymers-Opportunities and Risks II: Sustainability, Product Design and Processing*, Eyerer, P.; Weller, M.; Hübner, C.& Agnelli, J.A., (Ed), 120, Springer, ISBN 978-3-642-02796-3, NY, USA

Inal, O.; Kılıcarslan, M.; Arı, N. & Baykara, T. (2008). In Vitro and In Vivo Transdermal Studies of Atenolol Using Iontophoresis. *Acta Poloniae Pharmaceutica and Drug Research*, Vol.65, No.1, (January-February 2008), pp.29-36, ISSN 0001-6837

Irfani, G.; Sunilraj, R.; Tondare, A. & Shivanad, N. (2011). Design and Evaluation of Transdermal Drug Delivery System of Valsartan Using Glycerine as Plasticizer. *International Journal of Pharma Research and Development*, Vol.3, No.2, (April 2011), pp. 185-192, ISSN 0974-9446

Jadhav, R.T.; Kasture, P.V.; Gattani, S.G. & Surana, S.J. (2009). Formulation and Evaluation of Transdermal Films of Diclofenac Sodium. *International Journal of Pharmaceutical Technology Research*, Vol.1, No.4, (October-December 2009), pp. 1507-1511, ISSN 0974-4304

Karunakar, R.; Suresh, B.; Raju, J. & Veerareddy, P.R. (2010). Venlafaxine Hydrochloride Transdermal Patches: Effect of Hydrophilic and Hydorphobic Matrix on In Vitro Characteristics. *Latin American Journal of Pharmacy*, Vol.29, No.5, (August 2010), pp. 771-776, ISSN 0326-2383

Kumar, T.M.P.; Umesh, H.M.; Shivakumar, H.G.; Ravi, V. & Siddaramaiah, V. (2007). Feasibility of Polyvinyl Alcohol as a Transdermal Drug Delivery System for Terbutaline Sulphate. *Journal of Macromolecular Science*, Part A, Vol.44, No.6, pp. 583-589, ISSN 1060-1325

Limpongsa, E. & Umprayn, K. (2008). Preparation and Evaluation of Diltiazem Hydrochloride Diffusion Controlled Transdermal Delivery System. *American Association of Pharmaceutical Scientists Pharmaceutical Science and Technology*, Vol.9, No.2, (June 2008), pp. 464-470, ISSN 1530-9932

Lin, S.Y.; Lee, C.J. & Lin, Y.Y. (1991). The Effect of Plasticizers on Compatibility, Mechanical Properties and Adhesion Strength of Drug Free Eudragit E Films. *Pharmaceutical Research*, Vol.8, No.9, (September 1991), pp. 1137-1143, ISSN 0724-8741

Lin, S.Y.; Chen, K.S. & Rhun-Chu, L. (2000). Organic Esters of Plasticizers Affecting the Water Absorption, Adhesive Property, Glass Transition Temperature and Plasticizer Permanence of Eudragit Acrylic Films. *Journal of Controlled Release*, Vol.68, pp. 343-350, ISSN 0168-3659

Madishetti, S.K.; Palem, C.R.; Gannu, R.; Thatipamula, R.P.; Panakanti, P.K. & Yamsani, M.R. (2010). Development of Domperidone Bilayered Matrix Type Transdermal Patches: Physicochemical, In Vitro and Ex Vivo Characterization. *DARU Journal of Pharmaceutical Sciences*, Vol.18, No.3, pp.221-229, ISSN 1560-8115

Mamatha, T.; Venkateswara, R.J.; Mukkanti, K. & Ramesh, G. (2010). Development of Matrix Type Transdermal Patches of Lercanidipine Hydrochloride: Physicochemical and In Vitro Characterization. *DARU Journal of Faculty of Pharmacy*, Vol.18, No.1, pp.9-16, ISSN 2008-2231

Marzouk, M.A.E.; Kassem, A.E.D.; Samy, A.M. & Amer, R.I. (2009). In Vitro Release, Thermodynamics and Pharmacodynamic Studies of Aceclofenac Transdermal Eudragit Patches. *Drug Invention Today*, Vol.1, No.1, (November 2009), pp. 16-22, ISSN 0975-7619

Meathrel WG. (2010). Challenges in Formulating Adhesives for Transdermal Drug Delivery. *Transdermal Magazine 9*, pp. 1-4.

Meier, M.M.; Kanis, L.A. & Soldi, V. (2004). Characterization and Drug Permeation Profiles of Microporous and Dense Cellulose Acetate Membranes: Influence of Plasticizer and Pore Forming Agent. *International Journal of Pharmaceutics*, Vol.278, pp.99-110, ISSN 0378-5173

Mukherjee, B.; Mahapatra, S.; Gupta, R.; Patra, B.; Tiwari, A. & Arora, P. (2005). A Comparison Between Povidone Ethyl Cellulose and Povidone Eudragit Transdermal Dexamethasone Matrix Patches Based on In Vitro Skin Permeation. *European Journal of Pharmaceutics and Biopharmaceutics*, Vol.59, pp. 475-483, ISSN 0939-6411

Mundada, A.S. & Avari, J.G. (2010). Evaluation of Gum Copal as Rate Controlling Membrane for Transdermal Application: Effect of Plasticizers. *Acta Pharmaceutica Sciencia*, Vol.52, pp. 31-38, ISSN 1307-2080

Murthy, S.N.; Hiremath, S.R.R. & Parajothy, K.L.K. (2004). Evaluation of Carboxymethyl Guar Films for the Formulation of Transdermal Therapeutic Systems. *International Journal of Pharmaceutics*, Vol.1, No.272, pp. 11-18, ISSN 0378-5173

Nesseem, D.I.; Eid, S.F. & El-Houseny, S.S. (2011). Development of Novel Transdermal Self Adhesive Films for Tenoxicam, an Anti-inflammatory Drug. *Life Sciences*, Vol.89, No.13-14, (September 2011), pp. 430-438, ISSN 0024-3205

Nicoli, S.; Colombo, P. & Santi, P. (2005). Release and Permeation Kinetics of Caffeine From Bioadhesive Transdermal Films. *American Association of Pharmaceutical Scientists Pharmaceutical Science and Technology*, Vol.7, No.1, (September 2005), pp. 218-223, ISSN 1530-9932

Nicoli, S.; Penna, E.; Padula, C.; Colombo, P. & Santi, P. (2006). New Bioadhesive Transdermal Film Containing Oxybutynin: In Vitro Permeation Across Rabbit Ear Skin. *International Journal of Pharmaceutics*, Vol.325, No.1-2, (November 2006), pp.2-7, ISSN 0378-5173

Padula, C.; Colombo, G.; Nicoli, S.; Catellani, P.L.; Massimo, G. & Santi, P. (2003). Bioadhesive Film for the Transdermal Delivery of Lidocaine: In Vitro and In Vivo Behavior. *Journal of Controlled Release*, Vol.3, No.88, pp. 277-285, ISSN 0168-3659

Padula, C.; Nicoli, S.; Aversa, V.; Colombo, P.; Falson, F.; Pirot, F. & Santi, P. (2007). Bioadhesive Film for Dermal Drug Delivery. *European Journal of Dermatology*, Vol. 17, No.4, (July-August 2007), pp. 309-312, ISSN 1167-1122

Padula, C.; Nicoli, S.; Colombo, P. & Santi, P. (2010). Single Layer Transdermal Film Containing Lidocaine: Modulation of Drug Release. *European Journal of Pharmaceutics and Biopharmaceutics*, Vol.66, pp. 422-428, ISSN 0939-6411

Pandit, V.; Khanum, A.; Bhaskaran, S. & Banu, V. (2009). Formulation and Evaluation of Transdermal Films for the Treatment of Overactive Bladder. *International Journal of Pharmaceutical Technology Research*, Vol.1, No.3, (July-September 2009), pp. 799-804, ISSN 0974-4304

Parthasarathy, G.; Reddy, K.B. & Prasanth, V.V. (2011). Formulation and Characterization of Transdermal Patches of Naproxen with Various Polymers. *Pharmacie Globale*, Vol.2, No.6, (July 2011), pp. 1-3, ISSN 0976-8157

Patel, R.P. & Baria, A.H. (2011). Formulation and Evaluation Considerations of Transdermal Drug Delivery System. *International Journal of Pharmaceutical Research,* Vol. 3, No.1, (Nov 2011), pp. 1-9, ISSN 0975-2366

Petereit, H.U.; Assmus, M.; Beckert, T.; Bergmann, G. (2005)Binding Agent Which is Stable in Storage and Used for Pharmaceutical Applications. US20050019381A1

Quan, D. Passive Transdermal Drug Delivery Systems (TDDS): Challenges and Potential. *Transdermal Magazine,* 3 (2011): 6-12.

Rahman, M. & Brazel, C.S. (2004). The Plasticizer Market: An Assessment of Traditional Plasticizers and Research Trends to Meet New Challenges. *Progress in Polymer Science*, Vol.29, pp.1223-1248, ISSN 0079-6700

Rajabalaya, R.; David, S.R.N.; Khanam, J. & Nanda, A. (2010). Studies on the Effect of Plasticizer on In Vitro Release and Ex Vivo Permeation From Eudragit E 100 Based Chlorpheniramine Maleate Matrix Type Transdermal Delivery System. *Journal of Excipients and Food Chemistry*, Vol.1, No.2, pp. 3-12, ISSN 2150-2668

Rajan, R.; Sheba Rani, N.D.; Kajal, G.; Sanjoy Kumar, D.; Jasmina, K. & Arunabha, N. (2010). Design and In Vitro Evaluation of Chlorpheniramine Maleate From Different Eudragit Based Matrix Patches: Effect of Plasticizer and Chemical Enhancers. *ARS Pharmaceutica*, Vol.50, No.4, pp. 177-194, ISSN 0004-2927

Rao, P.R. & Diwan, P.V. (1997). Permeability Studies of Cellulose Acetate Free Films for Transdermal Use: Influence of Plasticizers. *Pharmaceutica Acta Helvetiae*, Vol.72, No.1, pp.47-51, ISSN 0031-6865

Repka, M.A. & McGinity, J.W. (2001). Bioadhesive Properties of Hydroxypropylcellulose Topical Films Produced by Hot-Melt Extrusion. *Journal of Controlled Release*, Vol.70, pp.341-351, ISSN 0168-3659

Salman N & Teutsch I. (2011). Transdermal Therapeutic System for Administration of Fentanyl or an Anologue Thereof. US 20110111013A1

Schroeder, I.Z.; Franke, P.; Schaefer, U.F. & Lehr, C.M. (2007a). Delivery of Ethinylestradiol From Film Forming Polymeric Solutions Across Human Epidermis In Vitro and In Vivo in Pigs. *Journal of Controlled Release*, Vol.118, pp. 196-203, ISSN 0168-3659

Schroeder, I.Z.; Franke, P.; Schaefer, U.F. & Lehr, C.M. (2007b). Development and Characterization of Film Forming Polymeric Solutions for Skin Drug Delivery. *European Journal of Pharmaceutics and Biopharmaceutics*, Vol.65, pp.111-121, ISSN 0939-6411

Selzer T. (2001). Transdermal Therapeutic System for Delivery of Dofetilide. US20010033859A1

Selzer T. (2004). Transdermal Therapeutic System for Releasing Venlafaxine. US20040101551A1

Shinde, A.; Shinde, A.L. & More, H. (2010). Design and Evaluation of Transdermal Drug Delivery System of Gliclazide. *Asian Journal of Pharmaceutics*, Vol.4, No.2, (April-June, 2010), pp. 121-129, ISSN 0973-8398

Siepmann, J.; Lecomte, F. & Bodmeier, R. (1999). Diffusion Controlled Drug Delivery Systems: Calculation of the Required Composition to Achieve Desired Release Profiles. *Journal of Controlled Release*, Vol.60, No.2-3, pp. 379-389, ISSN 0168-3659

Tanner, T. & Marks, R. (2008). Delivering Drugs by the Transdermal Route: Review and Comment. *Skin Research and Technology*, Vol. 14, No. 3, pp. 249-260, ISSN 0909-752X

Tiwary, A.K.; Sapra, B. & Jain, S. (2007). Innovations in Transdermal Drug Delivery: Formulations and Techniques. *Recent Patents on Drug Delivery and Formulation*, Vol. 1, No. 1, pp. 23-36, ISSN 1872-2113

Vasil'ev, A.E.; Krasnyuk, I.I. & Tokhmakhchi V.N. (2001). Transdermal Therapeutic Systems for Controlled Drug Release. *Pharm. Chem. J.*, Vol. 35, pp. 613-626, ISSN 1573-9031

Walters, K.A. & Brain, K.R. (2002). Dermatological Formulation and Transdermal Systems. In: *Dermatological and Transdermal Formulations*. Walters KA (Ed), 319-400, Informa, 319-400, ISBN 0-8247-9889-9, NY, USA

Wang, C.C.; Zhang, G.; Shah, N.H.; Infeld, M.H.; Malick, A.W. & McGinity, JW. (1997). Influence of Plasticizers on the Mechanical Properties of Pellets Containing Eudragit RS 30 D. *International Journal of Pharmaceutics*, Vol.152, pp.153-163, ISSN 0378-5173

Wang, F.J.; Yang, Y.Y.; Zhang, X.Z.; Zhu, X.; Chung, T. & Moochala, S. (2002). Cellulose Acetate Membranes for Transdermal Delivery of Scopolamine Base. *Materials Science and Engineering*, Vol.20, No.2, pp. 93-100, ISSN 0928-4931

Whelan, T. (1994). *Polymer Technology Dictionary*, Chapman&Hall, 308-309, ISBN 0-412-58180-9, London,UK

Williams, A. (2003). *Transdermal and Topical Drug Delivery: From Theory to Clinical Practice.* Pharmaceutical Press, 169-194, ISBN 0-85369-489-3, London, UK.

Wu, C. & McGinity, J.W. (1999).Non Traditional Plasticization of Polymeric Films. *International Journal of Pharmaceutics*, Vol.177, pp. 15-27, ISSN 0378-5173

Wypch, G. (2004). *Handbook of Plasticizers*, Chem Tec, 437-440, ISBN 1-895198-29-1, Ontario, Canada

Zhu, Y.; Shah, N.H.; Malick, A.W.; Infeld, M.H. & McGinity, JW. (2002). Solid State Plasticization of an Acrylic Polymer With Chlorpheniramine Maleate and Triethyl Citrate. *International Journal of Pharmaceutics*, Vol.241, pp.301-310, ISSN 0378-5173

6

Plasticizers and Their Role in Membrane Selective Electrodes

Mohsen M. Zareh*
Department of Chemistry, Faculty of Science,
Zagazig University, Zagazig
Egypt

1. Introduction

In the last four decades, the uses and application of ion-selective electrodes was widely applied in several researches as well as analytical projects. These electrodes are varied between solid state electrodes, liquid membrane electrodes, gas membrane electrodes, and plastic membrane electrodes. Plasticizers are one of the major components of the plastic membranes. As it is usually known, they are responsible of their physical properties.

In the beginning, a general idea about ion-selective membrane electrodes (Janata *Principles of Chemical Sensors* 1989) will be useful to understand the role of plasticizers as they are part of the constituents for some of them. These membranes are the main component of the potentiometric ion sensors. They are responsible of forming a type of discrimination in the electrode behavior towards one ion rather than others. A potential difference will be aroused when the analyte ion can penetrate across the phase boundary between the two phases (analyte solution, and internal reference solution). Accordingly, an electrochemical equilibrium will be formed, due to different potentials at both sides of the membrane. The potential difference (E) across the membrane is described by the Nernst equation:

$$\mathbf{E = E^\circ + (RT/ZF)\ ln\ a} \quad (1)$$

Where, E° : is the standard cell potential, R: general gas constant, Z: valency of the analyte ion, F:Faraday's constant, a: activity of the analyte ion.

If the activity of the target ion at side A is kept constant, the unknown activity at side B ($a_A = a_x$) is related to (E) by the following equation (at equilibrium condition):

$$E = RT/Z_xF \cdot \ln (a_x/a_A) = \text{const} + S \cdot \log (a_x) \quad (2)$$

where S=59.16/Z [mV] at T=298 K.

There are two main groups of membranes, which are used in ion-selective electrodes: namely, crystalline and noncrystalline membranes (Buck and E. Lindner, *Pure & App.Chem.* (1994).

*Corresponding Author

The crystalline membrane electrodes might be homogeneous or heterogenous. The homogenous type is prepared from single compound or a homogenous mixture (e.g. Ag_2S, AgI/Ag_2S). An example is the fluoride selective electrode based on LaF_3 crystals.

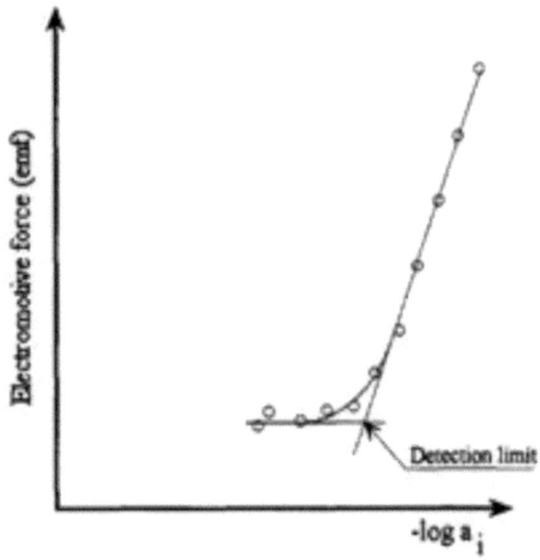

Fig. 1. Relation between EMF of the cell and activity of the analyte.

On the other side, the heterogenous membrane electrodes are formed by mixing an active substance or a mixture of active substances with an inert matrix (e.g. silicon rubber, PVC, hydophobized substances or conducting epoxy.

Non-crystalline membrane electrodes are composed of a support matrix, containing ion-exchanger (either cationic or anionic), a plasticizer solvent, and possibly an uncharged selectivity-enhancing species as a membrane which interposed between two aqueous solutions. The support matrix might be macroporous (poly-propylene carbonate, glass frit) or microporous (thirsty glass or inert polymeric material such as PVC)

1.1 Construction of the electrodes

These electrodes are prepared from glass capillary tubing approximately 2 millimeters in diameter, a large batch at a time. Polyvinyl chloride is dissolved in a solvent and plasticizers (typically phthalates) added, in the standard fashion used when making something out of vinyl. In order to provide the ionic specificity, a specific ion channel or carrier is added to the solution; this allows the ion to pass through the vinyl, which prevents the passage of other ions and water.

The measurements procedures using such electrodes are based on connecting the electrode to the terminal of a galvanometer or pH meter (G), the other terminal is connected to a reference electrode. Then, both electrodes are immersed in the solution to be tested. The concentration of the tested solution can be estimated from the galvanometer reading by the aid of the previously constructed calibration graph (for standard analtye concentrations).

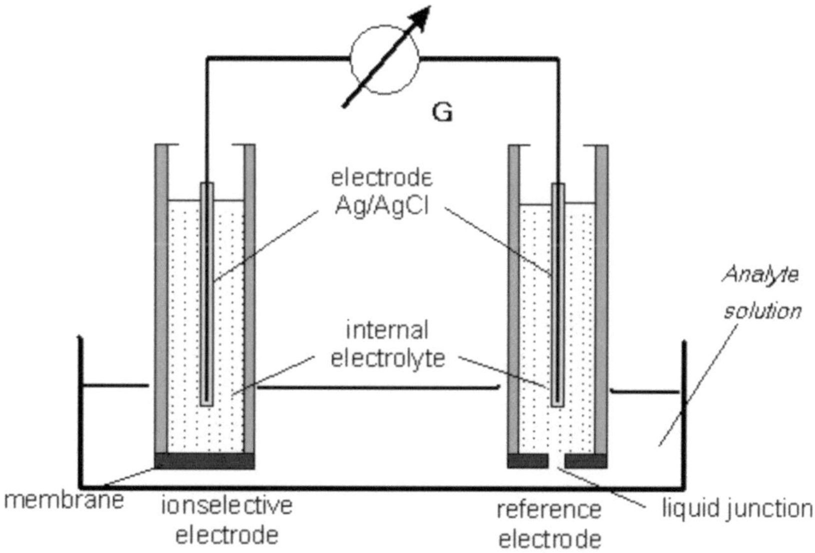

Fig. 2. Arrangement of a cell used for measurements by ion-selective electrodes.

1.2 Plasticizers for membrane electrodes

PVC-membranes for the selective electrodes are actually plastics. Different types of plasticizers are usually applied for preparing the membrane electrodes. Many types of plasticizers are used for plastic membranes as additives. Phthalates are considered as the most common types which produce the desired flexibility and durability. They are esters of polycarboxylic acids with either linear or branched aliphatic alcohol. The plasicizer embeddy itself between the chains of the polymer. It spaces them apart, so the free volume increases. The more the added plasticizer, the more flexiblility and durability (lower cold flex temperature).They might be ethers, esters of either aromatic or aliphatic acids. The ester plasticizer can be classified into dicarboxylic/tricarboxylic ester-based plasticizers.

Several plasticizers are recorded for preparing ion-selective membrane electrodes (Zareh, *Sensor Letters*, 2010). Some common phthalate plasticizers are [Chanda and Roy (2006). Plastics technology handbook]. This type of plasticizer is commonly applied for preparing membrnaes of the ion-selective electrodes. Structure of selected plasticizers commonly applied for plastic membranes are shown in figure 3.

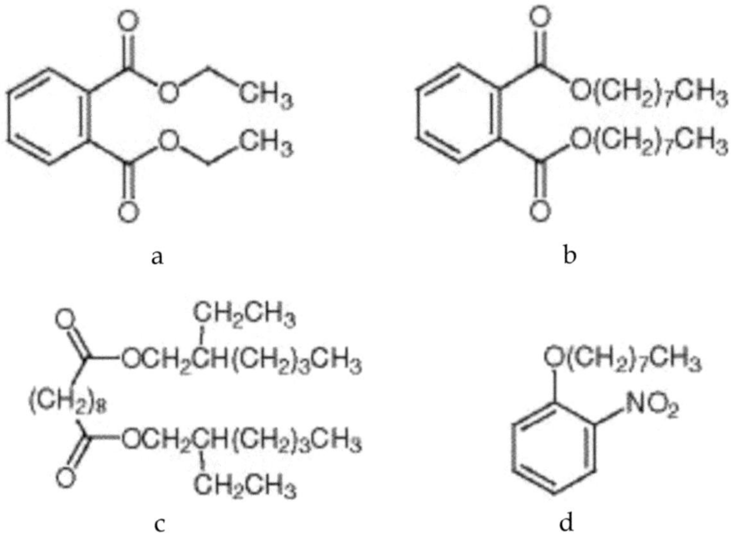

Fig. 3. Structure of selected plasticizers used for preparing ion selective electrode membranes:a) Diethyl phthalate, b) Di-n-octyl phthalate, c) bis(2-ethylhexyl) sebacate, d) 2-Nitrophenyl n-octyl ether.

No.	Plasticizer	Ion response	Reference
1	Dioctyl phthalte, didecylphthalate, dioctyl sebacate	Ascorbic acid,	Zareh, *Sensor Letters.*, 2010
2	Nitophenyl octyl ether, didecylphthalate, dioctyl sebacate	Quinine	Zareh et al, *Analyt.Chim. Acta.*, 2001
3	Dioctyl phthalte dioctyl sebacate	Surfactant	Espadas-Torre, *Anal.Chem.* 1996
4	Dioctyl phenyl phosphonate	Pb^{++}	Quagraine and Gadzekpo, *Analyst.*, 1992
5	Dibutyl phthalate	nitrate	Péreza et al, *Sens. and Act. B*, 2003
6	Dioctyl sebacate	Organic amines	Odashima et al, *Anal. Chem.*,1993
7	Dioctyl phenyl phosphonate	Fe^{+++}	Zareh et al, *Electroanalysis*, 2010
8	tetrakis (2-ethylhexyl) pyromellitate	Cocaine	Watanabe et al, *Analytica Chimica Acta 1995*
9	Dibutyl phthalate, acetophenone, dimethyle sebacate	Triiodide	Khayatian et al, *Anal. Sci.* 21, 297-302

Table 1. Selected examples of several plasticizers used in preparing ion selective electrodes.

From the above mentioned plasticizers, phthalates (especially dioctylphthalate, diethyl phthalate, dibutyl sebacate, dinitrophenyl octyl ether, didecyl phtalate are widely used in preparing PVC-membranes for the ion-selective electrodes.

2. How can a plasticizer work like an ionophore?

Usually, plastic membranes of ion-sensitive electrodes are composed of sensing material (ionophore, or ion-exchanger), PVC, and plasticizers (Zamani, *Materials Science and Engineering: C* 2008; Ekmekci, *Journal of Membrane Science*,2007). The basic requirements of adequate plasticizer are four criteria. The plasticizer must exhibit sufficient lipophibicity, no crystallization in the membrane and no oxidation. In addition, it must fulfill the selectivity properties (Eugster et al, *Analyt. Chim. Acta*, 1994). They tried to built plasticizers to fulfill all the requirements. They can not found a relation between selectivity coefficient and the dielectric constant of the plasticizer. From the obtained results they concluded that the selectivity properties was improved in presence of plasticizers without functional groups which can compete with the carriers.

Different plasticizers were applied like an ionophore as mentioned before. The sensing materials were considered for long time to be responsible of the selectivity properties of the membrane electrodes. Blank membranes (ionophore -free) were tried for measurements of H^+ (Zareh, *Analytical Sciences* 2009). The electrodes of this type were working Nernstainly. This is due to that the plasticizer with a donor site can work like ion exchangers for cationic species like H^+.

The effect of plasticizer can be clearly found, if only membranes without an ionophore were involved. An important role of plasticizers was recently discovered by (Zareh, *Analytical Sciences* 2009). In that work electrodes IIIa, IIIb, and IIIc with blank membranes were prepared. They contain only plasticizers NPOE, DOS, or DDP into the PVC matrix without any ionophore. Figure 4 shows the calibration graphs for the blank membrane electrodes (IIIa - IIIc), when H^+ was measured in H_2SO_4. The NPOE membrane IIIa showed the best Nernstian response, 57.17 mV/decade. The other membranes (IIIb and IIIc) deviated from the Nernstian behavior. The functional groups and the donation sites of the plasticizers are proved to affect the chelation of the primary ion. In DOS and DDP, the ester groups are the main part in the molecule. This group is usually inactive regarding the coordination interactions. Due to the lone pair of the oxygen atom, the ether group in NPOE is more likely to be associated with the H^+. Therefore, this membrane exhibits Nernstian response towards H^+. In the absence of the ionophores, the plasticizer link to the primary ion and perform the ion exchange process, which will lead to the potential variation. The equilibrium can be represented as:

$$K_{eq} = [\text{NPOE H}^+]^+{}_m / [\text{H}^+]_s [\text{NPOE}]_m$$

$$(O_2N\text{-}C_6H_4\text{-}O\text{-}C_8H_{17})_m + (H^+)_s = ([O_2N\text{-}C_6H_4\text{-}O\text{-}C_8H_{17}]\,H^+)^+{}_m$$

where "m": refers to membrane site, and "s": refers to the solution site.

Tohda et al (*J. Mol. Str.*, **1997**) applied the Second Harmonic Generation (SHG)-technique for membranes without an ionophore, but in absence of the primary ion. They tried a PVC-membrane with DOS as a plasticizer. They reported that the SHG-signal is neglected for the DOS-membrane electrode without ionophore. This agrees with the poor Nernstian response for such membrane in the present study. They did not studied the SHG-signals for membranes without ionophores for NPOE nor for DDP-membrane electrodes.

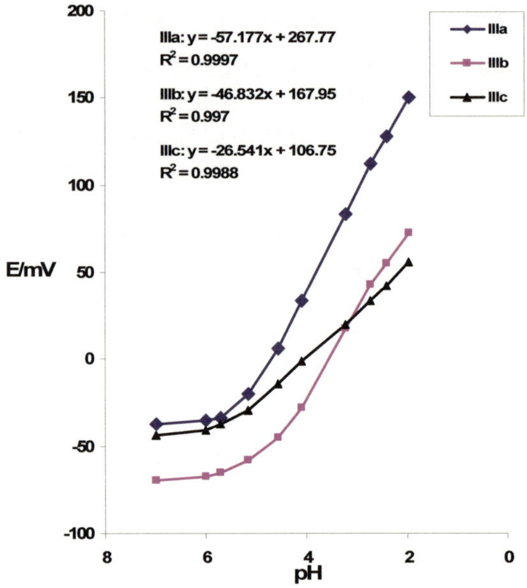

Fig. 4. Calibration graphs for blank membranes in presence of different plasticizers. (Zareh, *Analytical Sciences*, 2009).

The behaviour of membranes containing an ionophore (N,Ń-bisethoxycarbonyl-1,10-diaza-4,7,13,16-tetraoxacyclo-octadecane (diaza-18-crown-6) (DZCE), 37,40-bis-[(diethoxy-thiophosphoryl)oxy]-5,11,17,23,29,35-hexakis(1,1-dimethylethyl)-calix[6]arene-8,39,41,42-tetrol (CAX), was tested to compare them with those given by blank membrane electrodes. Figure 5, shows the results.

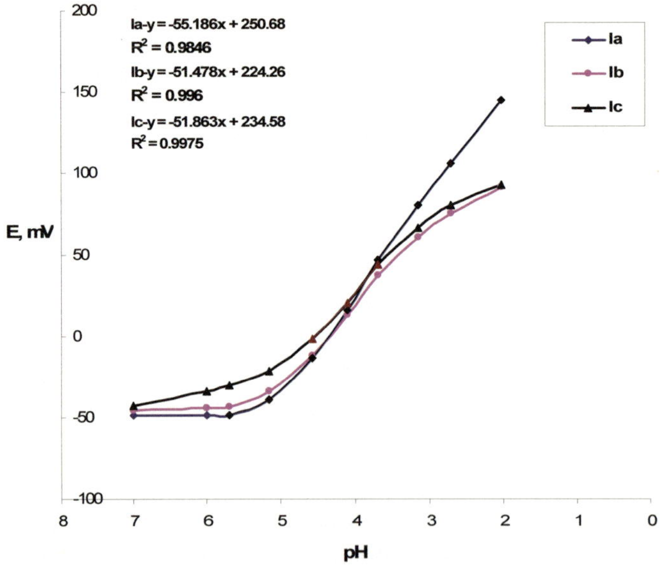

Fig. 5. Effect of plasticizer on the electrode performance based on DZCE-ionophore (Zareh, *Analytical Sciences*, 2009).

The calibration graphs of the blank electrode IIIa in the presence of 0.1M NaCl (to adjust ionic strength), showed a drop in the slope value 17.7 mV/decade. This drop was not observed for electrode Ia under the same condition. So, it can be concluded that the presence of an ionophore enables the membrane to carry the primary ion and subsequently stabilizes the electrode behavior (slope, linearity, response time, and selectivity). Figure 6 shows the calibration graphs of the tested electrodes. In addition, the presence of the ionophore prolonged the electrode age (from 1 week for IIIa to 4 weeks for Ia).

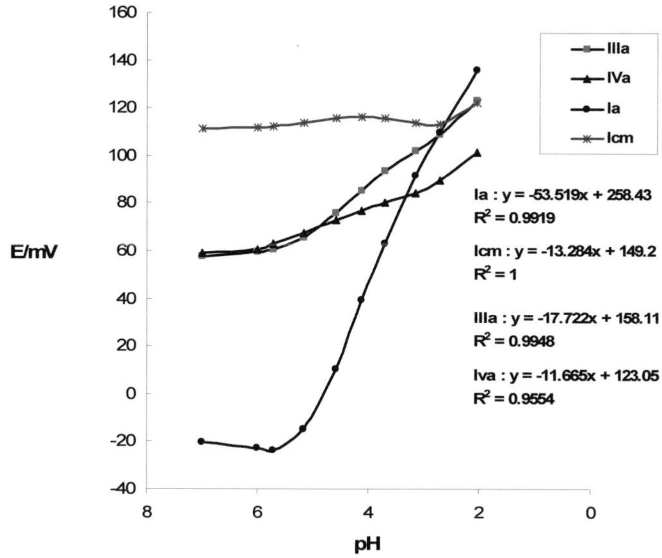

Fig. 6. Effect of the presence of 0.1MNaCl on the performance of different electrodes (Zareh, *Analytical Sciences*, 2009).

3. Plasticizer and selectivity properties

Selectivity properties is one of the most important properties of an ion-selective electrode. The evaluation of the selectivity is a major criteria upon which the electrode is considered either selective or not. There are several methods for determining the selectivity coefficient values ($K_{A,B}$ ^{pot}). These can be mentioned below according to the IUPAC definition (Buck and Lindner, *Pure& Applied Chem, 1994)*:

3.1 Fixed Interference Method (FIM)

The emf of a cell comprising an ion-selective electrode and a reference electrode **(ISE** cell) is measured with solutions of constant activity of interfering ion, a_B, and varying activity of the primary ion. The emf values obtained are plotted *vs.* the logarithm of the activity of the primary ion a_A. The intersection of the extrapolation of the linear portions of this plot indicates the value of a_A which is to be used to calculate ($K_{A,B}$ ^{pot}) from the Nikolsky-Eisenman equation A,B:

$$K_{A,B}^{pot} = \frac{a_A}{a_B^{z_A/z_B}}$$

3.2 Separate Solution Method (SSM)

The emf of a cell comprising an ion-selective electrode and a reference electrode (ISE cell) is measured with each of two separate solutions, one containing the ion A of the activity a_A (but no B), the other containing the ion B at the same activity $a_B = a_A$ (but no A). If the measured values are E_A and E_B, respectively, the value of may be calculated from the equation:

$$\log K_{A,B}^{pot} = \frac{(E_B - E_A)z_A F}{2.303RT} + \left(1 - \frac{z_A}{z_B}\right)\lg a_A$$

3.3 The Separate Solution Method SSM) II

The concentrations of a cell comprising an ion selective electrode and a reference electrode (ISE cell) are adjusted with each of two separate solutions, one containing the ion A of the activity a_A (but no B), the other containing the ion B (but no A) of the activity as high as required to achieve the same measured cell voltage. From any pair of activities a_A and a_B giving the same cell voltage, the value of ($K_{A,B}^{pot}$) may be calculated from the

$$K_{A,B}^{pot} = \frac{a_A}{a^{z_A/z_B}}$$

The FIM and SSM methods are recommended only when the electrode exhibits a Nernstian response to both principal and interfering ions. These methods are based on the assumption that plots of E_I *vs.* $\lg(a_A^{1/|z_A|})$ and E_2 *vs.* $\lg(a_B^{1/|z_B|})$ will be parallel and the vertical spacing is $(2.303RT/F)\lg K_{A,B}^{pot}$.

However, the FIM can always be used to determine a minimum primary ion concentration level at which the effect at interference can be neglected. The actual conditions of the FIM method match the conditions under which the electrodes are used.

What is interesting is that these electrodes (ionophore-free) showed significant selectivity properties. This is found from the values of the selectivity coefficient. The selectivity coefficient values of an electrode ($K_{A,B}^{pot}$) is calculated by the SSM for different common cations. Table (2) shows an example of the obtained results. The values were compared to those for electrodes containing ionophoric sensing material (diaza-18-crown-6 for Ib, Ic; and phosphorylated calix-6-arene for IIa). In case of other electrodes, (Ib, Ic, and IIa), high selectivity coefficient values were observed. These values indicate that the selectivity properties of electrodes Ib, Ic and IIa are lower than those for Ia electrode. This is attributed to the chelating property of the plasticizer NPOE. To prove this effect of the plasticizer, the selectivity coefficient values for the blank membrane electrodes IIIa, IIIb and IIIc were calculated. Likewise Ia, the selectivity coefficient values of IIIa electrode is better than those for IIIb and IIIc electrodes. The second important observation from the selectivity coefficient values of blank electrodes was that these values are very close to the values recorded for electrodes Ia, Ib, and Ic. Accordingly, the plasticizer plays an important role in the selectivity properties of the membrane electrodes.

Interferent	Ia	Ib	Ic	IIa	IIIa	IIIb	IIIc
Na^+	1.2×10^{-3}	1.5×10^{-2}	3.4×10^{-2}	1.3×10^{-1}	5.7×10^{-4}	5.5×10^{-2}	3.4×10^{-1}
K^+	2.7×10^{-3}	5.2×10^{-2}	1.2×10^{-1}	1.7×10^{-1}	1.1×10^{-3}	3.2×10^{-1}	1.7
Cs^+	2.4×10^{-3}	5.1×10^{-2}	1.1×10^{-1}	1.8×10^{-1}	1.2×10^{-3}	2.8×10^{-1}	2.4
NH_4^+	2.2×10^{-3}	3.1×10^{-2}	5.4×10^{-2}	1.8×10^{-1}	9.8×10^{-4}	1.3×10^{-1}	6.6×10^{-1}
Mg^{++}	1.3×10^{-4}	2.3×10^{-3}	8.0×10^{-3}	1.6×10^{-2}	8.7×10^{-5}	1.7×10^{-3}	1.4×10^{-2}
Ca^{++}	9.2×10^{-5}	2.0×10^{-3}	4.8×10^{-3}	1.7×10^{-2}	6.4×10^{-5}	1.1×10^{-3}	8.1×10^{-3}
Ba^{++}	8.3×10^{-5}	1.8×10^{-3}	9.8×10^{-3}	1.8×10^{-2}	4.9×10^{-5}	2.2×10^{-3}	2.8×10^{-2}
Pb^{++}	2.0×10^{-3}	8.7×10^{-2}	2.0×10^{-1}	3.2×10^{-2}	1.3×10^{-3}	3.0×10^{-1}	5.0×10^{-1}
Zn^{++}	1.7×10^{-4}	2.4×10^{-2}	5.4×10^{-3}	1.7×10^{-2}	8.0×10^{-5}	1.5×10^{-3}	1.0×10^{-2}

Table 2. Selectivity coefficient values ($K_{H^+, J^{z+}}$) for H^+-electrodes based on diazacrown ether analogues (I), phosphorated calix[6]arene (II), and blank membranes (III).

4. Plasticizer and detection limit

Plasticizers affect the detection limit of the different type of electrodes. **Bedlechowicz et al, 2002** *Journal of Electroanalytical Chemistry,* studied the effect of the plasticizer on the extended linear calibration curve and on the selectivity of a calcium selective electrode with ETH 1001 ionophore as a function of calcium activity in the internal solution. (2-Ethylhexyl)sebacate (DOS) and *o*-nitrophenyloctyl ether (*o*-NPOE) were used as plasticizers. The poly(vinylchloride) membrane also contained potassium tetrakis(4-chlorophenyl)borate. The linear part of the calibration curve of the electrode with *o*-NPOE is longer and the detection limit is lower compared to values for the electrode containing DOS as the plasticizer. The optimal activity of free Ca^{2+} and Na^+ in the internal reference solution was 10^{-4} and 10^{-1} for the membrane with DOS and 10^{-6} and 10^{-1} for the membrane with *o*-NPOE, respectively. The repeatability of the response for electrodes with the lowest detection limit is similar in the case of both plasticizers. The selectivity coefficients were determined for electrodes having activities of calcium ion in the internal solution in the range from 10^{-2} to 10^{-10}. The properties of the electrodes can be correlated with the transport properties of their membranes.

The same conclusion was recorded by **Gupta et al 2000,** *Talanta,* for Cd electrode. They studied the potential response of cadmium(II) ion selective electrode based on cyanocopolymer matrices and 8-hydroxyquinoline as ionophore has been evaluated by varying the amount of ionophore, plasticizer and the molecular weight of the cyanocopolymer. The sensitivity, working range, response time, and metal ions interference have shown a significant dependence on the concentration of ionophore, plasticizer and molecular weight of cyanocopolymers. The electrodes prepared with 2.38×10^{-2} mol kg^{-1} of ionophore, 1.23×10^{-2} mol dm^{-3} of plasticizer and 2.0 g of cyanocopolymer (molecular wt., 59 365) have shown a Nernstian slope of 29.00±0.001 mV per decade activities of Cd^{2+} ions with a response time of 12±0.007 s. Electrodes have shown an appreciable selectivity for Cd^{2+} ions in the presence of alkali and alkaline earth metal ions and could be used in a pH range of 2.5–6.5. The cyano groups of the copolymers contributed significantly to enhance the selectivity of the electrode. The electrode has shown an appreciable average life of 6 months without any significant drift in the electrode potential and found to be free from leaching of membrane ingredients. Electrode response is explained considering phase boundary model based on thermodynamic considerations.

5. Plasticizer and di-electric constant

The plasticizer showed an effect on the dielectric constant of a memebrane. **Kumar and Sekhon,** *European Polymer Journal,* **2002**, studied such effect by addition of plasticizer to the polyethylene oxide (PEO)–ammonium fluoride (NH_4F) polymer electrolytes. They found to result in an increase in conductivity value and the magnitude of increase has been found to depend upon the dielectric constant of the plasticizer. The addition of dimethylacetamide as a plasticizer with dielectric constant (ϵ =37.8) higher than that of PEO (ϵ =5) results in an increase of conductivity by more than three orders of magnitude whereas the addition of diethylcarbonate as a plasticizer with dielectric constant (ϵ =2.82) lower than that of PEO does not enhance the conductivity of PEO–NH_4F polymer electrolytes. The increase in conductivity has further been found to depend upon the concentration of plasticizer, the concentration of salt in the polymer electrolyte as well as on the dielectric constant value of the plasticizer used. The conductivity modification with the addition of plasticizer has been explained on the basis of dissociation of ion aggregates formed in PEO–NH_4F polymer electrolytes at higher salt concentrations.

Similar study was applied earlier by **William Robert**, *Ploymer* **1998**. They studied the influence of plasticizer on the dielectric characteristics of highly plasticized PVC.Three citrate-related compounds [Citroflex A-4 (CFA4), Citroflex A-6 (CFA6), and Citroflex B-6 (CFB6)] and six sebacate-related compounds [dimethyl sebacate (DMS), diethyl sebacate (DES), dibutyl sebacate (DBS), dioctyl sebacate (DOS), dioctyl azelate (DOZ), and dioctyl adipate (DOA)] were used to evaluate the effects of configurational changes in plasticizer on the dielectric properties of ion-selective poly(vinyl chloride) membranes. Tridodecylamine (TDDA) and potassium tetrakis-4-chlorophenyl borate (KTpClPB) were used as neutral charge carriers and negative sites, respectively. Using parallel plate sensors, the dielectric properties [ionic conductivity (σ) and tan δ] of the plasticized PVC membranes were determined at temperatures from − 100 to + 100°C and seven log frequencies (− 1, 0, 1, 2, 3, 4, and 5 Hz). Generally, increasing the amount of plasticizer in the membrane improved the σ and lowered the temperature of the tan δ peak. A positive linear correlation existed between the log σ and the log phr ratio for a given temperature and frequency, when no data was included for membranes below the melting temperature of the plasticizer. When plotted *versus* temperature, the slopes of all these lines passed through a maximum between 0 and 60°C. The intercepts of all these lines increased monotonically with increasing temperature. These intercepts were highly dependent on the frequency at low temperatures, becoming less frequency dependent as the temperature increased. Having established that configurational changes of the plasticizers had no effect above the melting point of each plasticizer, global nomograms were only required for the citrate- and sebacate-related plasticizers, respectively. Using the appropriate nomogram for a selected plasticizer, the σ could be predicted at a given phr ratio, temperature, and frequency.

6. Plasticizers and the physical properties of membrane

Puncture tests quantified five mechanical properties for at least eight levels of seven plasticizers were applied (**Gibbons et al**, *Polymer*, **1997**). Using Citroflex B-6 at a phr ratio of 0.31, the strength and secant stiffness peaked at 9.63 N and 1250 N m^{-1}, respectively. At a phr ratio of 0.6 the toughness peaked at 48 N mm. These three properties decreased at higher phr ratios for all plasticizers. Tangent stiffnesses were generally 1.7 times secant stiffnesses. They concluded that for all plasticizers, ductility increased to a constant value of 15 mm at a phr ratio of two. The molecular structures of the plasticizers influenced the mechanical properties. For a given phr ratio, plasticizers having lower hydrodynamic volumes increased the strengths, stiffnesses, and toughnesses of the membranes. Compared to prior dielectric testing, the strength, toughness, and stiffness increased as the ionic resistivity increased. In electrodes and biosensors phr ratios should be reduced to a minimum of one.

7. Plasticizers and health

Many vinyl products contain additional chemicals to change the chemical consistency of the product. Some of these additional chemicals called additives can leach out of vinyl products. Plasticizers that must be added to make PVC flexible have been additives of particular concern. The leaching out of plasticizers during measurements and their volatility (during preparing membranes) are the main causes of the toxicity by the plasticizers.

8. References

Buck R.P. and Lindner E., Pure & App. Chem. 1994, 66, 2527–2536.

J. Janata, *Principles of Chemical Sensors*, 1989, Plenum Press, New York,

Zareh M., *Sensor Letters*, 2010, 8, 1-8.

Chanda, Manas; Roy, Salil K. (2006). *Plastics technology handbook*. CRC Press. pp. 1–6.

http://www.tciamerica.com/catalog/S0025.html.

Zamani, H. A., Hamed-Mosavian, M. T., Hamidfar, E., Ganjali, M. R., Norouzi, P., *Materials Science and Engineering: C* 2008, 28, 1551-1555.

Zareh M., Malonwska E., Kasiura K., *Analyt.Chim. Acta* 2001, 447, 55.

Espadas-Torre C., Bakker E., Barker S., and Meyerhoff M., *Anal.Chem.* 1996, 98, 1623-1631.

Quagraine E. and Gadzekpo V., *Analyst*, 1992, 117,1899-1903.

Péreza M., Marínb L. P., Quintanab J., Yazdani-Pedramc M., *Sensors and Actuators B: Chemical*, 2003, 89, 262–268.

Odashima K., Yagi K., Tohda K., and Umezawa Y., *Anal. Chem.*, 1993, 55, 1074-1083

Zareh M., Ismail I., and Abd El-Aziz M., *Electroanalysis* 2010, 22, 1369-1375.

Watanabe K., Okada K., Oda *H.*, Furuno K., Gomita Y., Katsu T., *Analytica Chimica Acta* 1995, 316, 371-375.

Khayatian G., Rezatabar H., Salimi A., *Anal. Sci.* 2005, 21, 297-302

Ekmekci, G., Uzun, D., Somer, G., Kalaycı, Ş., *Journal of Membrane Science* 2007, 288, 36-40.

Eugster R., Rosatin T., Rusterholz B., Aebersold B., Pedrazza U., Ruegg D., Schmid A., Spichiger U.E., Simon W., *Analyt. Chim. Acta*, 1994, 289, 1-13.

Gupta V., Jain A., Agarwal S., Maheshwari G., *Talanta* 2007, 71, 1964-1968.

Sil A., Ijeri V., Srivastava A., *Sensors and Actuators B: Chemical* 2005, 106, 648-653.

Mashhadizadeh, M. H., Shoaei, I. S., Monadi, N., *Talanta* 2004,

Zareh M., *Anal.Sci.* 25, 1131(2009).

K. Tohda, S. Yoshiyagawab, Y. Umezawaa, *J. Mol. Str.*, 1997, 408/409, 155.

Buck R., and Lindner E., *Pure and Applied Chem* 1994, 12, 2527-2538.

Bedlechowicz I., Maj-Żurawska M., Sokalski T., Hulanicki A., *Journal of Electroanal. Chem.* 2002, 537, 111-118.

Gupta K.C, Jeanne D'Arc M., *Talanta*,2000, 52, 1087-1103

Kumar M., Sekhon S.S. *European Polymer Journal*, 2002,38, 1297-130.

William S., Gibbons, and Robert P. Kusy *1998, Polymer, 39, 3167-3178*

Gibbons W., Patel H., Kusy R., *Polymer*, 1997, 38, 2633-2642

7

Use of Plasticizers for Electrochemical Sensors

Cristina Mihali and Nora Vaum
North University of Baia Mare
Romania

1. Introduction

Plasticizers represent a category of organic substances that can be added to polymers in order to improve some of their characteristics like elasticity and mechanical properties. Plasticizers are very important for the design of the selective polymeric membrane for potentiometric sensors, an important class of the electrochemical sensors. They fix the dielectric constant of the membrane and improve the selectivity of those devices. Ion - selective electrodes are instruments used in the potentiometric analysis. They have analytical applications in fields such as environment analysis, pharmaceutical analysis and, quality assurance in different fabrications. Choosing an appropriate plasticizer as well as an adequate electrode component and sometimes an additive is essential for the construction of an ion- selective electrode with high analytical performances such as sensibility, selectivity, fast response, and long lifetime.

2. Plasticizers used for the polymeric membranes of ion-selective electrodes

The development of plasticized polymeric membrane sensors was a big step forward. It led to the advance and diversification of ion-selective electrode analysis. The liquid membranes of the electrodes were difficult to handle and did not allow the use of ion-selective electrodes (ISEs) in any position because the liquid membrane would leak. The polymeric membrane has properties similar to those of liquid membranes, but the range of applications is much larger. Plasticized polymers are in fact highly viscous liquids and they are known in the literature as liquid membranes due to high values of diffusion coefficients of ionophores and their complexes. This membrane can still be considered as a liquid phase, because diffusion coefficients for a dissolved low-molecular-mass component (e.g., an ionophore) are on the order of 10^{-7} to 10^{-8} cm^2 s^{-1} (Moody & Thomas, 1979 as cited in Oesch et al., 1986). Typically, such a solvent polymeric membrane contains about 66 g of plasticizer and only 33 g of PVC per 100 g. Only at very low plasticizer contents (<20 g/100 g), diffusion coefficients may be 10^{-11} cm^2 s^{-1} and smaller, approaching values that are found for solids.

The plasticizers used in the preparation of the polymeric membrane of ion -selective electrodes must be compatible with the polymer and electrodic component and also must be solved in tetrahydrofuran or cyclohexanone, the solvent used in the membrane preparation. The plasticizers with high lipophilicity are preferred. The most used plasticizers are: *ortho*-nitrophenyloctyl ether (NPOE), dibutyl phthalate (DBP), dinonyladipinate (DNA), tris(2-ethylhexyl) phosphate (TEHP), tris(ethylhexyl) phosphate (TEHP), bis(2-ethylhexyl) adipate (DOA), dioctylphthalate (DOP), and bis (2-ethylhexyl) sebacate (DOS).

This way a large constructive variety of polymeric membrane sensors could be made with or without an internal reference solution, including the sensors used in flow injection analysis. The systematization of the bibliographic material and our own research material was pursued in order to point out the way in which the properties of the plasticizers influence the characteristics of the potentiometric sensors (whose main components are ion-selective electrodes) and the general way of properly selecting a suitable plasticizer. Plasticizers used for the preparation of polymeric membranes used in the construction of ion-selective electrodes for inorganic and organic ions are presented. We also presented the performances obtained as well as how to select a plasticizer for the construction of ion-selective electrodes used in pharmaceutical products, anionic surfactants, physiologically active amines and inorganic ions analysis.

2.1 The membrane potential and the function of the ion–selective electrodes

A potentiometric sensor with polymeric membrane is shown in Figure 1.

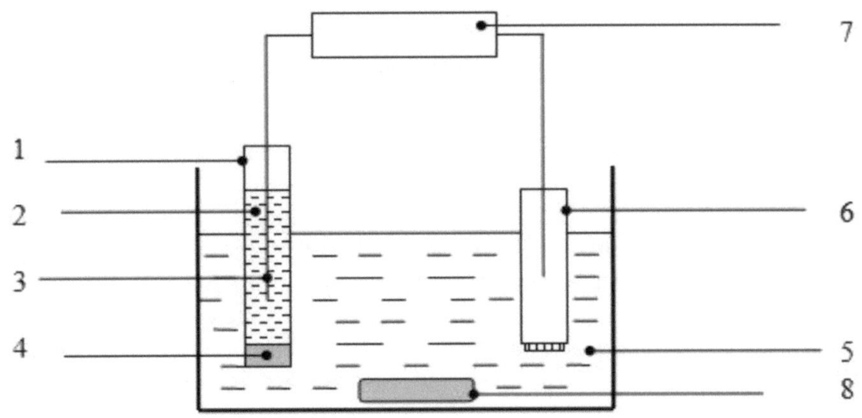

Fig. 1. Potentiometric sensor (1 – ion-selective electrode; 2 – internal reference solution; 3 – internal reference electrode; 4 – membrane; 5 – test solution\sample; 6 – external reference electrode; 7 – milivoltmeter; 8 – magnet piece for magnetic stirring).

Figure 2 shows the contact phases and electrochemical potentials occurring in an electrochemical cell used for potentiometric measurements.

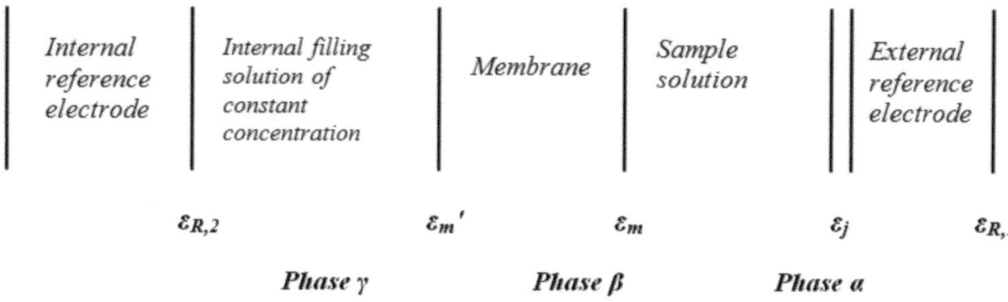

Fig. 2. Electrochemical cell for potentiometric analysis.

The membrane potential ε_m, for an ion *i* with a z_i charge, is expressed by the equation (1):

$$\varepsilon m = \frac{RT}{ziF}\ln\frac{(ai)_\alpha}{(ai)_\beta} \quad (1)$$

In which R represents the ideal gas constant, T represents the absolute temperature, F represents the Faraday constant. $(a_i)_\alpha$ represents the activity of the primary ion *i* in phase α, $(a_i)_\beta$ represents the activity of the primary ion *i* in the phase β.

Indexes α and β refer to the two phases: α - the analyte and β - the membrane. Phase γ consists of an aqueous solution of known concentration that contains an existent ion and phase β (membrane). If this ion is ion *i* the electrochemical cell is a concentration cell and the membrane potential can be considered a concentration potential. The membrane potential ε_m' which appears at the contact of phase β and γ will have a constant value because the common ions activity is constant in both phases. Potentials ε_m and ε_m' do not appear as a result of oxidation or reduction, but due to some ion exchange equilibriums in which the analyzed species participate. The difference in potential E_c between the two electrodes of the electrochemical cell, represented in figure 2, is given by the following equation:

$$E_c = \varepsilon_m + \varepsilon_{R,1} + \varepsilon_m' - \varepsilon_{R,2} - \varepsilon_j \quad (2)$$

ε_j is the junction potential. If $\varepsilon_{R,1}$, $\varepsilon_{R,2}$, ε_m' and ε_j values are considered constant equation (2) can be written:

$$Ec = const. + \varepsilon m = const. + \frac{RT}{ziF}\ln\frac{(ai)_\alpha}{(ai)_\beta} \quad (3)$$

If the activity of ion *i* in the membrane $(a_i)_\beta$ is constant the membrane potential ε_m varies by the *i* species activity, according to a Nernstian law:

$$Ec = const. + \frac{RT}{ziF}\ln(ai)_\alpha \quad (4)$$

Considering a working temperature of 25°C, equation (4) becomes:

$$E_c = const. + \frac{0,059}{z_i}\log(a_i)_\alpha \quad (5)$$

In fact, the strict obedience of this law by a potentiometric sensor is disturbed by some interferences. They appear because of the fact that the membranes are not perfectly selective and are permeable to ions other than the primary ion.

$$E_c = const. \pm \frac{RT}{zF}\ln\left(a_i + \sum K_{ij}a_j^{Zi/Zj}\right) \quad (6)$$

in which j represents the interfering species, a_j the activity of the interfering ion, and z_j it's charge. k_{ij} is the potentiometric selectivity coefficient. The sign of the logarithmic term is "+" if i is a cation and "-" if it is an anion.

2.2 The preparation of plasticized polymeric membranes

The preparation of plasticized membranes is relatively simple. It can be used to construct a great variety of polymeric membrane sensors selective to many inorganic and organic ions. The polymeric membrane contains the following substances: the electrodic component, the plasticizer, and the polymeric substance. The electrodic component can be an organic ion exchanger called ionophore, a neutral sequestrant or a complex combination. This component makes the membrane sensitive to the species that needs to be analyzed because it is responsible for the appearance of the membrane potential due to the repartition equilibrium between the sample and membrane phases. The analyte in the membrane phase is involved in a chemical equilibrium with the ionophore.

The polymeric substance is usually polyvinyl chloride (PVC) with high molecular mass, but other polymers are also used (polyurethane and polyaniline). Plasticizers are used in a relatively large proportion, generally 66%. The plasticizer assures the mobility of the ion exchanger, fixes the dielectric constant value of the membrane and confers it the adequate mechanical properties. Choosing a suitable plasticizer is important because it improves the ion-selective electrodes sensitivity. During the time, plasticizer is gradually released from the polymeric membrane due to the contact of the analyzed solution with the water that enters the membrane. The membrane becomes opaque. Membrane components are dissolved in a suitable solvent (tetrahydrofuran or cyclohexanone) and they mix with the formation of a viscous liquid that is poured on a flat surface and left to dry slowly for 48 hours in a solvent vapor (tetrahydrofuran or cyclohexanone) saturated atmosphere. By evaporation a thin polymer film is formed. The polymeric membrane is then cut into disk forms and glued to an electrode body in which the internal reference solution and the internal reference electrode are introduced. Another approach consists in the deposit of the membrane on a metallic pill made out of Ag or Cu or on a graphite rod. The electrodes that have been prepared this way are left covered by a glass bell to avoid the rapid evaporation of the solvent which may affect the homogeneity of the membrane. Figure 3 shows a classic membrane electrode with an internal reference solution.

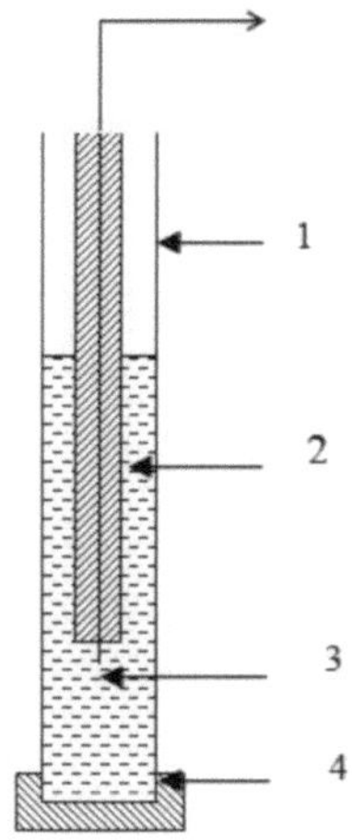

Fig. 3. Polymeric membrane electrode with an internal reference solution. 1 - Electrode body; 2 - Internal reference electrode (usually Ag - AgCl); 3 - Internal electrolyte; 4 - Plasticized polymeric membrane.

Figure 4 shows a "coated wire" ion-selective electrode.

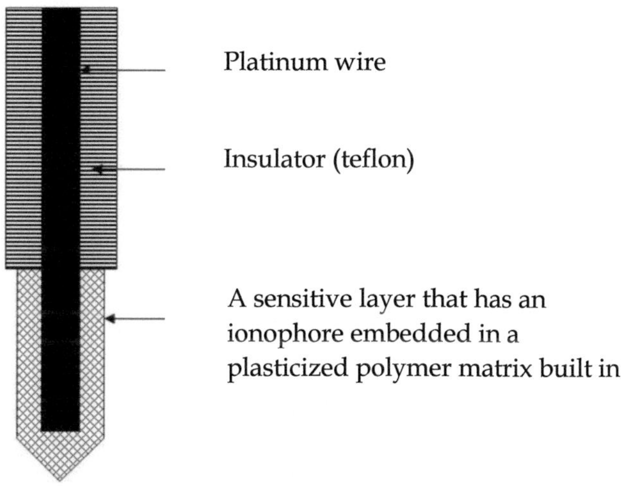

Fig. 4. Ion-selective electrode with a "coated wire" polymeric membrane.

The electrode coating mixtures containing the plasticizer, PVC of high molecular mass and the ionophore or the ion-exchange sensing material is obtained by dissolving 1 g of mixture in 20 cm^3 of tetrahydrofuran. A metallic wire (Ag, Pt, Cu, Al) or a teflonised graphite electrode which served as a membrane carrier is dipped in the coating mixture and after evaporation of the solvent, the procedure must be repeated twice.

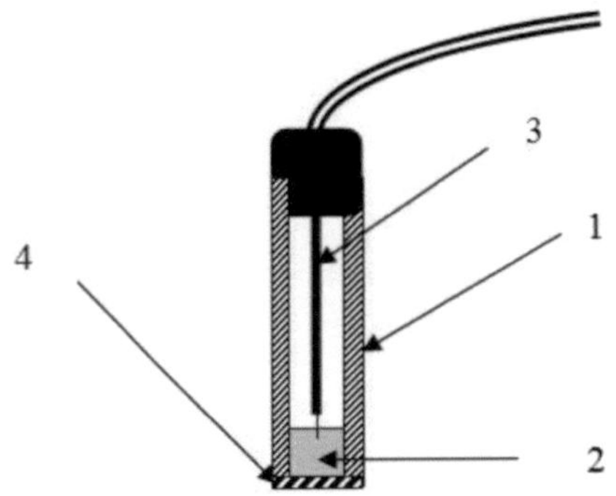

Fig. 5. The design of an ion - selective electrode with a PVC membrane deposed on a metallic pill 1 - PVC body; 2 - Cu or Ag pill; 3 - coaxial cable; 4 - PVC membrane sensible to a certain anion or cation.

The time of usage for a polymeric membrane electrode is limited to 1 to 6 months. This aspect is compensated by the low price and simple manufacturing (Mihali et al., 2008).

2.3 The selection of a proper plasticizer

The desirable properties of a plasticizer used in the membrane preparation of the ion-selective electrodes are: compatibility with the polymer, low volatility and low solubility in aqueous solution, low viscosity, low cost and low toxicity (O'Rourke et al., 2011). In order to select the best plasticizer usually some tests are necessary. Electrodes with different compositions are built, in which both the nature of the ionophore and of the plasticizers are modified and their proportions in the membranes are changed. The properties of the ion-selective electrodes with different membrane compositions are tested. The electrode which has the proper characteristics is selected. The most important considered characteristics are: linear response range, slope (sensitivity), and also selectivity towards the ions that can be present in the analyzed solution. An electrode used to determine species i can respond to species j. The selectivity coefficient shows the electrode sensitivity ratio for different species. A low value of the selectivity coefficient shows a low interference toward certain chemical specie.

3. Selecting plasticizers for inorganic ion sensitive potentiometric sensors

New ion-selective electrodes for potassium were developed and tested employing 18-crown-6-ether, dibenzo-18-crown-6-ether, and 4',4''(5'')-di-tert-butyldibenzo-18-crown-6-ether (dbdb-18-6) ionophores in PVC membranes with a polyaniline solid contact between the membranes and the Pt substrate (Han et al., 2008). Many types of sensors have been developed to measure the concentration of potassium ions. Various plasticizers (ortho-nitrophenyloctyl ether (NPOE), tris(ethylhexyl) phosphate (TEHP), bis(2-ethylhexyl) adipate (DOA), dioctylphthalate (DOP), and bis (2-ethylhexyl) sebacate (DOS)) were tested for the best response. It should be remembered that the nature of the plasticizer influences both the dielectric constant of the membrane and the mobility of the ionophore and its complex. With the 18-6 ionophore, DOP and NPOE produced better results than other plasticizers in response slope (RS), but DOS and DOA yielded the best detection limit (DL). With the dibenzo-18-6, DOP and DOS yielded a comparably better RS, but NPOE, DOS, and DOP yielded comparable DL. With the dbdb-18-6, DOA, TEHP and DOS gave better RS, and DOA and DOS yielded a better DL. Most of these had detection limit below 10^{-5} M and a response slope below 50 mV/decade, which are not better than the valinomycin-based ISEs. Only the electrode with the dibenzo-18-crown-6-ether ionophore with DOA as the plasticizer showed a better performance than the valinomycin ISEs: response slope of 58 mV/decade and DL of $10^{-5,82}$.

The performances of tetracycline based cation selective polymeric membrane electrodes of many sets with different plasticizers were investigated as the selectivity of ion-selective electrodes and optodes are greatly influenced by membrane solvent and also controlled by plasticizers. A membrane with bis(2-ethylhexyl) sebacate and additive shows good potentiometric performance toward Ca^{2+} (slope: 27.8 mV per decade; DL: −4.52) including selectivity (Baek et al., 2007). Contrastingly, a membrane with dibutyl phthalate shows near Nernstian response, it has also shown the best measuring range and detection limit for Ca^{2+} (29.5 mV and 5.10) and Mg^{2+} (24.4 mV and −5.04) and the least selectivity has been also observed between Ca^{2+} and Mg^{2+}. When both membranes were used together to flow system, the concentration of Ca^{2+} and Mg^{2+} could be determined, simultaneously.

Based on the concept of ion-selective conductometric micro sensors (ISCOM) a new calcium sensor was developed and characterized. Optimization of the membrane composition was carried out by testing different types of calcium-ionophores, polymers, and plasticizers. The most commonly used membrane material is based on plasticized high molecular PVC. In general, only a limited number of commercially available calcium ionophores and plasticizing agents are used in Ca^{2+}-ISE (Trebbe et al., 2001). The tested plasticizers are: 2-Nitrophenyl octyl ether (NPOE), dibutyl sebacate (DBS), dioctylphenylphosphonate (DOPP), bis(1-butylpentyl)decane-1,10-diyl diglutarate ETH 469 (BDD), 2-fluorophenyl-2-nitrophenylether (2F2NE), and tetrahydrofuran. In order to investigate the influence of membrane components 14 different membrane compositions containing different commercially available ionophores, plasticizers, and polymers have been tested. It was concluded that plasticizers used in organic solvent membranes have to fulfill many criteria, e.g. high lipophilicity, solubility in the polymeric membrane (no precipitations) as well as no exudation (one phase system) and a good selectivity of the resulting membrane. Moreover, with regard to the ISCOM operation mechanism, a high polarity may be advantageous in order to extract ionic species into the membrane phase. Membranes based on plasticizer (NPOE) with a quite high polarity and moderate lipophilicity showed the best properties.

Rapid and economical procedures for determining aluminum(III) in aqueous solutions are required in the industry of aluminum compounds. A plasticized Al-selective electrode was fabricated and studied. The composition of a membrane was as follows: the ionophore of an aluminon, 10 mM (~1 wt %); PVC, 66 wt %; dibutyl phthalate (DBP) as plasticizer, 33 wt %. The possibility of using the developed ISE for determining aluminum(III) in aqueous solutions was proved (Evsevleeva et al., 2005).

4. Selecting plasticizers for anionic surfactant sensitive potentiometric sensors

The surfactants are compounds essential to the modern civilization and technology. Determination and monitorization of the surfactants concentration is necessary in the production of detergents, in the industrial processes where anionic surfactants are used, in quality control of products containing added surfactants and in environment surveillance activities, especially monitoring water quality. In the last decades there have been created and improved numerous analytic methods for determination of anionic surfactants. Among these, there are the potentiometric methods based on potentiometric sensors (electrodes sensible to anionic surfactants). These are very attractive tools due to their good precision, relatively simple manufacturing, relatively low cost and their ability to determine the surfactants in the samples without previous separation steps (Mihali, 2006). The use of surfactant sensitive electrodes with plasticized polymeric membrane for the potentiometric determination of low concentrations of anionic surfactants has been described in several papers (Lizondo-Sabater et al, 2008; Matesic-Puac et. Al, 2005; Mihali et al., 2009; Nemma et al., 2009; Oprea et al. 2007 etc.)

The selection of the most suitable plasticizer for the ion selective electrodes sensible to lauril sulfate anion based on ion-association complexes of quaternary ammonium cation and surfactant anion was made by testing the behavior of poly(vinyl chloride) membrane composition with two ionophores (Mihali et al., 2009) : cetyltrimethylammonium laurylsulfate (CTMA-LS) and tricaprylmethylammonium laurylsulfate (TCMA-LS). As plasticizers were used tricresylphosphate (TCF), ortho-nitrophenyloctylether (NPOE) and dioctylsebaccate (DOS). Comparing the electrodes prepared from the point of view of the linear response range and the slope (Table 1) we can notice that the best performances have been obtained with the CTMA-LS ionophore based electrode, plasticized with DOS (slope 59.39 mV/concentration decade, linear response range 10^{-3}-$3.93x10^{-6}$M) and the electrode with TCMA-LS ionophore and the same plasticizer (slope: 58.56 mV/concentration decade, linear response range 10^{-3}-$2x10^{-6}$). Nearer values have been obtained for membrane with TCMA-LS ionophore, plasticized with TCF (slope: 58.89 mV/decade and linear response range 10^{-3}-$2.9x10^{-6}$ M).

Ionophore / Plasticizer		Membrane characteristics	
		Slope, mV/conc. decade	Linear response range, M
Cetyltrimethylammonium laurylsulfate (CTMA-LS)	DOS	59.39	10^{-3}-$3.93x10^{-6}$
	TCF	58.19	10^{-3}-$4.2x10^{-6}$
	NPOE	56.08	10^{-3}-$4.88x10^{-6}$
Tricaprylmethylammonium laurylsulfate (TCMA-LS)	DOS	58.56	10^{-3}-$2x10^{-6}$
	TCF	58.87	10^{-3}-$2.9x10^{-6}$
	NPOE	55.07	10^{-3}-$3.5x10^{-6}$

Table 1. Influence of polymeric membrane composition on the performances of laurilsulfate sensible electrode.

The effect of different plasticizers in the sensing membrane on the performance of a surfactant ISE based on a PVC membrane with no added ion-exchanger was investigated. *o*-nitrophenyl octyl ether (NPOE), o-nitrophenyldecyl ether (NPDE), *o*-nitrophenyl dodecyl ether (NPDOE) and *o*-nitrophenyl tetradecyl ether (NPTE) were used as plasticizers. Electrodes based on NPDE, NPDOE and NPTE produced better results than NPOE-plasticized PVC membrane electrodes in terms of low detection limits. Electrodes based on NPDE, NPDOE and NPTE displayed a Nernstian slope in the concentration range of 10^{-6} to 10^{-2} M. NPOE plasticized PVC membrane electrodes displayed a Nernstian slope in the concentration range of 10^{-5} to 10^{-2} M (Masadome et al., 2004). The three electrodes other than the NPOE - plasticized PVC membrane electrode showed a similar performance to that of the NPOE - plasticized PVC membrane electrode concerning low detection limits and slope sensitivity. The four electrodes examined in this study are excellently selective for the dodecyltrimethylammonium ion over inorganic anions, but interference from other cationic surfactants such as tetradecyltrimethylammonium ions is significant. With respect to slope sensitivity, selectivity, response time and pH effect, the four electrodes showed a similar performance. The use of NPOE derivatives as the more hydrophobic plasticizers in comparison to NPOE enhanced the performance of surfactant-selective PVC membrane electrodes with no added ion-exchanger with respect to low detection limits. On the other hand, the use of NPOE derivatives did not enhance the performance of surfactant - selective PVC membrane electrodes with respect to slope sensitivity, selectivity, response time or pH effect.

The cyclam derivative 1,4,8,11-tetra(*n*-octyl)-1,4,8,11-tetraazacyclotetradecane (L) has been used as carrier for the preparation of PVC-based membrane ion-selective electrodes for anionic surfactants (Lizondo-Sabater et al, 2008). Different membranes were prepared using L as ionophore, tetra *n*-octylammonium bromide (TOAB) as cationic additive and dibutyl phthalate (DBP) or *o*-nitrophenyl octyl ether (NPOE) as plasticizers. The final used electrode contained a membrane of the following composition: 56% DBP, 3.4% ionophore, 3.8% TOAB and 36.8% PVC. This electrode displays a Nernstian slope of -60.0 ± 0.9 mV/decade in a 2.0×10^{-3} to 7.9×10^{-6} mol dm^{-3} concentration range and a limit of detection of 4.0×10^{-6} mol dm^{-3}. The electrode can be used for 144 days without showing significant changes in the value of slope or working range. The electrode shows a selective response to dodecyl sulfate (DS^-) and a poor response to common inorganic cations and anions. The selective sequence found was $DS^- > ClO_4^- > HCO_3^- > SCN^- > NO_3^- \approx CH_3COO^- \approx I^- > Cl^- > Br^- > IO_3^- \approx NO_2^- \approx SO_3^{2-} > HPO_4^{2-} > C_2O_4^{2-} > SO_4^{2-}$, i.e. basically following the Hoffmeister series except for the hydrophilic anion bicarbonate. Most of the potentiometric coefficients determined are relatively low indicating that common anions would not interfere in the DS^- determination. A complete study of the response of the electrode to a family of surfactant was also carried out. The electrode showed a clear anionic response to DS− and to Na-LAS (sodium alkylbezenesulfonate) and a much poorer response to other anionic surfactants and to non-ionic surfactants. Also the electrode shows certain non-linear cationic response in the presence of cationic and zwitterionic surfactants. The electrode was used for the determination of anionic surfactants in several mixtures, and the results obtained were compared to those found using a commercially available sensor.

5. Selecting plasticizers for the construction of ion-selective electrodes used in pharmaceutical products analysis

Potentiometry with ion-selective electrodes is one of the most useful analytical tools capable of measuring both inorganic and organic substances in pharmaceutical products (Kulapina & Barinova, 1997). At present, most works are devoted to establishing the factors providing desired modification of the properties of ion-selective membranes (Kulapina & Barinova, 1997). In this context, interesting results were reported on the effects of the ion association and the character of a plasticizing solvent on the selective properties of counter electrodes for the organic cations of drugs. It was suggested that the factor of ion association can be used to obtain electrodes with increased selectivity. An exemplary case study includes the proserine-selective electrode, in which substitution of the NPOE plasticizer, having high dielectric constant with the lower dielectric constant DBP lead to a significant increase in selectivity towards tertiary ammonium cations in the presence of amine and metallic cations. Researches on the dimedrol-selective electrode showed the same influence of the plasticizer's dielectrical properties on the electrode selectivity with respect to physiologically active amines. Using plasticizers with high basicity, as in the case of trihexylphosphate, leads to increases in selectivity with respect to cations containing electrophilic hydrogen atoms. The donor-acceptor interaction of the non-ionic polar groups of amines with the solvation active centers of the membranes macrocomponents is responsible for this high increase in selectivity.

Many pharmaceutical products belonging to different categories (such as analgesics and anesthetics, antibiotics, vitamins, neuroleptics, antiseptics, antiviral preparations, etc.) contain an amino functional group, in their structure, more or less substituted or a nitrogen atom included in a heterocyclic compound. They are called physiologically active amines (PPA) (Egorov et al, 2010). The ISEs selectivity towards the cations formed by the PPA can be controlled by changing the nature of the plasticizer, the ion exchanger or by introducing a neutral carrier capable of selectively complexing the analyte ions in the membranes. The most used method, and probably, the most effective method is the variation of the plasticizer's nature. The influence of the plasticizer's nature upon its selectivity towards PPAs was investigated (Egorov et al, 2010). Different PVC membrane compositions and ISEs sensible to PPAs were prepared and studied. Tetrakis (A-chlorophenyl potassium)borate (TCPB) and potassium tris(2,3,4-nonyloxy)benzene sulfonate were used as ion exchangers. In some membrane compositions a neutral carrier: dibenzo-18-crown-6 (DB-18-C-6) or a solvating additive: mono-decylresorcinol (MDR) was used. The influence of the plasticizer's basicity upon the selectivity of the ISEs was studied and therefor the following plasticizers were used: *ortho*-nitrophenyloctyl ether (NPOE), dibutyl phthalate (DBP), dinonyladipinate (DNA), tris(2-ethylhexyl) phosphate (TEHP), tris(ethylhexyl) phosphate 2 (TEHP), bis(2-ethylhexyl) adipate (DOA), dioctylphthalate (DOP), and bis (2-ethylhexyl) sebacate (DOS). The basicity increases as the series: NPOE<DBP<DNA<TEHP<DOPP. It was established that by increasing the plasticizer's basicity there is a pronounced increase in membrane selectivity towards PPA cations with a low substitution of the salt forming nitrogen atom. PPA atoms that contain non-ionic polar groups and can form hydrogen bonds with Lewis bases behave in the same way, but PPA cations forming intermolecular hydrogen bonds are an exception (Egorov et al, 2010; Kulapina & Barinova, 1997). The effect of the ion exchanger's nature upon PPA cation selectivity was stronger when plasticizers with low basicity were used. It was established that the selectivity of the membranes which were prepared using highly basic plasticizers does not depend on the ion exchanger's nature. The ISEs that contain the neutral carrier dibenzo-18-crown-6 (DB-18-C-6) have shown the best selectivity towards primary amines cations in comparison with secondary and tertiary, when low basicity plasticizers were used. It was found that the use of high basicity plasticizers leads to a significant increase of the ISE selectivity towards primary amines cations in comparison to quaternary amines.

The plasticizer nature can significantly affect the selectivity of an ISE reversible to amine cations only if at least one of two conditions is met: if the analyte and foreign ions differ in the degree of substitution of the salt-forming nitrogen atom or if they differ in nature and (or) by the number of nonionic polar groups, capable of specifically interacting with the plasticizer, primarily, in the mechanism of the formation of hydrogen bonds. PAA cations, as a rule, possess sufficiently complex structures and, along with the salt-forming nitrogen atom, contain nonionic polar groups capable of specific solvation, which prevents the estimation of the "pure" effect of the plasticizer due to the difference in the degrees of substitution of the corresponding PAA cations. The selectivity coefficients of a Bu_4N^+-SE to the cations of primary, secondary, and tertiary amines strongly depend on the plasticizer nature and increases in the series NPOE < DBP < DNA < TEHP < DOPP. It was established that with an increase in the basicity of the plasticizer in the series NPOE < DNA < TEHP, the selectivity coefficients to quarternary ammonium cations naturally decrease, and the selectivity coefficients to the cations of secondary and, particularly, primary aliphatic amines increase.

The selectivity coefficients to the cations of triethyl and tributylammonium, and also to the cations of Dimedrol, Spasmolytin, Papaverin, Ketoconazole, and Vinpocetine slightly depend on the plasticizer nature. This results from the fact that, despite the strong difference in the structures of the above amines, their molecules, as the molecule of Ganglerone contain no groups capable of forming hydrogen bonds with basic plasticizers. On the contrary, the selectivity coefficients of the Ganglerone-SE to the cations of Bromhexine, Quinine, Trimecaine, Novocaine, and Pyridoxine significantly increase with an increase in the basicity of the plasticizer because of the presence of nonionic polar groups, -C(O)-NH-, -NH_2, -OH capable of forming hydrogen bonds with basic plasticizers in their molecules. The strongest effect is observed on the pyridoxine cation, with three hydroxyl groups. The cations of Loperamide and Metoclopromide take exception to that general regularity due to the formation of intermolecular hydrogen bond. Thus, despite the presence of polar groups in their molecules capable of forming hydrogen bonds with Lewis bases the change of the basicity of the plasticizer slightly affects selectivity to the Loperamide cation while for the Metoclopromide cation, the effect is opposite to that expected: selectivity coefficient decreases from NPOE to TEHP.

A diclofenac-selective electrode for ionometric analysis was developed using polyvinyl chloride membranes plasticized with dioctylphthalate (DOP), dibutylphthalate (DBP), dinonylphthalate (DNP), dinonylsebacenate (DNS), tricresylphosphtate (TCP), in which the electrode-active substance was an ionic associate of diclofenac and neutral red. The electrode contained an ionic associate of diclofenac with neutral red and its response was linear over the diclofenac concentration range 5×10^{-5} to 5×10^{-2} M with an electrode function slope of (30.0 ± 1.1) - (44.0 ± 1.2) mV/pC, that differed according to the plasticizers that were used. This membrane electrode was used as a sensor for assaying diclofenac in pharmaceutical formulations (Kormosh et al., 2009).

Procedures for determining ibuprofen with ion - selective electrodes are characterized by their rapidity, simplicity and low cost of the equipment, possibility of analyzing turbid and colored solutions, and acceptable selectivity and sensitivity (Nazarov et al., 2010). Membranes were prepared from polyvinil chloride (PVC), tetrahydrofuran (THF), and also a plasticizer dibutyl phthalate (DBP).

Currently, medicinal substances can be determined with a wide range of ion-selective electrodes of different types (biospecific, ion-exchange, neutral-carrier, etc.) with different electroanalytical characteristics (Kulapina & Barinova, 2001).. Tetrahydrofuran was used as the solvent for polyvinyl chloride; as solvent–plasticizers, dibutyl phthalate, dioctyl phthalate, and tributyl phosphate of analytical grade and *o*-nitrophenyloctyl ether were used. Solvents were purified by distillation; refractive indices were measured on a laboratory refractometer. It is known that electroanalytical properties of plasticized membrane electrodes are determined by the properties of both the ion exchanger and the solvent plasticizer. Therefore, electrode and transport properties of membranes were studied based on the above ion pairs. Transport properties of membranes were studied by the electrical conduction method and under applied potential conditions. The introduction of the compound of the medicinal substance and tetraphenyl borate into a membrane increases its conductivity (0.4 - 3.1 μS at $C_{ionophore}$ = 0.001 – 0.1 mol/kg of dibutyl phthalate) compared to the background conductivity (0.25 μS). Specific conductivity of membranes is nearly unaffected by the nature of the counterion; this indicates that mobilities of cations of medicinal substances in the membrane phase are close to each other. The apparent dissociation constant of the novocaine tetraphenyl borate ion pair in the membrane phase was estimated from stationary values of specific conductivity at 1.6×10^{-2}.

From the total of the obtained data, it follows that membranes under study (based on medicinal substance–tetraphenyl borate ion pairs) belong to liquid membranes with a dissociated ion exchanger, whose selectivity is controlled predominantly by the solvating ability of the solvent–plasticizer with respect to the analyte ion (Kulapina & Barinova, 2001). This fact was confirmed by complex studies on revealing the influence of the nature and concentration of the ionophore, the nature of the solvent plasticizer (dibutyl phthalate, dioctyl phthalate, tributyl phosphate, or *orto*-nitrophenyloctyl ether), and the polyvinyl chloride–plasticizer ratio in the membrane on the selective properties of electrodes. The comparative analysis of the electrochemical behavior of membranes was based on the results of studying the main operation parameters of ion-selective electrodes based on some membranes. It was demonstrated that electrodes with blank membranes containing polyvinyl chloride and the above solvent plasticizers exhibit the cationic function in solutions of physiologically active amines, and in this case, the linearity range and the slope of electrode functions depend on the nature of the plasticizer. Membranes based on tributyl phosphate lose elasticity after operation for two weeks because of a significant solubility of tributyl phosphate in water. Electrodes based on dibutyl phthalate and dioctyl phthalate exhibit the best electrode characteristics. The introduction of an ionophore significantly improves electrochemical and operating characteristics of electrodes. Similarly to blank membranes, electrodes based on dibutyl phthalate (as well as dioctyl phthalate) exhibit the best characteristics, i.e., the selectivity of ion-exchange membranes is determined by the solvating ability of the plasticizer. The nature of the medicinal substance incorporated into the ionophore only insignificantly affects the linearity range of electrode functions.The experimental data on varying the composition of membranes demonstrated that membranes based of dibutyl phthalate (as well as dioctyl phthalate) with a polyvinyl chloride-plasticizer ratio of 1 : 3 and the concentration of the ionophore 0.01 mol/kg of plasticizer exhibit the optimal characteristics.

The influence of the polarity and chemical structure of the plasticizer on the potentiometric response of the electrodes was investigated. Comparing the results, the best electrodes, containing methyltrioctylammonium chloride (MTOA–Cl) in the membrane, are appropriately plasticized with DBP, NPOE and TEHP (Lenik et al., 2008). For electrodes containing tetraoctylammonium chloride (TOA-Cl) in the membrane, similar potentiometric responses were obtained. It can be concluded that the chemical structure, lipophilicity and polarity of the plasticizers did not exert any considerable influence on the parameters of calibration graphs, either with tetraoctylammonium- or methyltrioctylammonium anion exchangers. The type of quaternary ammonium salt and plasticizer affects the selectivity coefficient of the naproxen electrodes. However, the polarity of plasticizers was not significant, resulting in the lowest selectivity towards lipophilic ions for the more hydrophilic salts, and the lower concentration of cationic sites in the membrane in consequence of the exchanger leaving the membrane.

An ion selective electrode for a vasoactive drug, Udenafil was developed (Han et al., 2010). Undafill is (5-[2-propyloxy-5-(1-methyl-2-pyrollidinylethylamidosulphonyl)phenyl]-1-methyl-3-propyl-1,6-dihydro-7-H-pyrazolo(4,3-d)pyrimidin-7-one). The electrode was based on ion pairs of Udenafil with different anions: tetrakis-*m*-chlorophenyl borate (TmClPB), tetraphenylborate (TPB) and phosphomolybdic anion (PMA). The slopes of electromagnetic field responses and the response range of a solid contact electrode based on Udenafil - TmClPB ion pair with those based on Udenafil - PMA and Udenafil - TPB ion pairs were compared, and showed that the response slopes were influenced by plasticizers. When electrodes with 6 different plasticizers based on Udenafi - TmClPB were compared, as the dielectric constant of PVC plasticizer increased, so was the response slope at the same time. Various plasticizers (DOS, NPOE, DOP, TEHP, DOA, DBP) were tested for best response (Han et al., 2010). An anionic additive KTpClPB (potassium tetrakis(4-chlorophenyl) borate) was also used. The composition of this solid contact electrode based on Udenafil - TmClPB ionophore was Udenafil - TmClPB ion pair 0.050 : PVC 0.190 : NPOE 0.350 : KTpClPB 0.001 (potassium tetrakis(4-chlorophenyl) borate) as additive. However, it appeared that remaining two electrodes compared to the electrode based on Udenafil - TmClPB ionophore had decreased Nernstian slopes and reduced response ranges (Han et al., 2010). However, as dielectric constant increased, so was the Nernstian slope. The solid contact electrodes based on Udenafil - TmClPB which used NPOE plasticizer (dielectric constant of about 24) showed the best Nernstian slope, and those using another plasticizer as DBP (6.4), DOP (5.1), DOS (4.6), DOA (about 4.0), TEHP (about 4.0) had lower Nernstian slope and narrower dynamic range. According to the increase in the dielectric constant of plasticizers, the Nernstian slope of solid contact electrode increases. The tendency of this type of electrodes is that the dielectric constant of plasticizer and liphophilicity of KTpClPB in PVC layer reduces the membrane resistance by reducing the activation barrier at the PVC outer surface sample solution interface and increases the mobility of Udenafil between the PVC outer surface and the sample solution. So, solid contact electrodes based on Udenafil - TmClPB plasticized with NPOE containing KTpClPB showed better recovery precision, response time and less standard deviation than solid contact electrodes with other plasticizers or with no KTpClPB at all.

A novel electrochemical sensor has been developed for the determination of nimesulide (NIM). The sensor is based on the NIM- molybdophosphoric acid (MPA) as the electroactive material in PVC matrix in presence of bis(2-ethyl hexyl) phthalate (BEP) as a plasticizer (Kumar et al., 2007).

Five different plasticizers were employed to study their effect on the electrochemical behaviour of the electrochemical sensor for nimesulfide: bis(2-ethyl hexyl) phthalate (BEP), bis(2-ethy hexyl) sebacate (BES), di-buthyl sebacate (DBS), bis(2-ethyl hexyl) adipate (BEA), di-n-butyl phthalate (DBP). Generally, the use of plasticizers improve certain characteristics of the membranes, and in some cases, the slopes get affected adversely. Here, the slopes in the case of the BES, DBS and DBP are super-Nernstian and BEA is sub-Nernstian. It was found that BEP gave a near Nernstian slope. The potentiometric response characteristics of the NIM sensor based on the use of NIM - MPA (ion association nimesulfide - molybdophosphoric acid). ion pair as the electroactive material and BEP as a plasticizer in a PVC matrix were examined (Kumar et al., 2007). The slope was 55.6 mV/decade over the concentration range 1.0×10^{-6} – 1.0×10^{-2} M of NIM.

The most used plasticizers in the construction of ion selective electrodes are summarized in Table 2.

Name of the plasticizer used in the preparation of the polymeric membrane, their proportion (wt%)	Polymer	Ion toward which the membrane is / Electrodic component	Performances of the sensor	Ref.
bis(2-ethylhexyl) adipate (DOA), 64,2 %	PVC, poly-aniline	K/4',4''(5'')-di-tert-butyldibenzo-18-crown-6-ether	Slope of 58mV/pC, detection limit (DL) of $10^{-5.8}$ M	(Han et al., 2008)
dibutyl phthalate (DBP) as plasticizer, 33 %.	PVC	Al /aluminon	Slope of 20.7 mV/pC, Linear response range: 10^{-5}-1 M	(Evsevleeva et al., 2005)
bis(2-ethylhexyl) sebacate (DOS), 66%	PVC	Lauryl-sulfate/ tricapryl-methyl-ammonium laurylsulfate	Slope of 58.56 mV/pC, Linear response range: $2\text{x}10^{-6}$-10^{-3} M	(Mihali et al., 2009)
orto-Nitrophenyl octyl ether (NPOE), 63 %	Poly-urethane	Ca/Calcium-ionophore I (ETH 1001)	Dynamic range of 10^{-6}-10^{-1} M, DL: 10^{-7}	(Trebbe et al., 2001)
dicotylphtalate (DOP), 66%	PVC	Diclofenac sodium/ion associate of diclofenac and neutral red	Slope of 44 mV/pC, Linear response range: $5\text{x}10^{-5}$ -$5\text{x}10^{-2}$ M	(Kormosh et al., 2009)
orto-Nitrophenyl octyl ether (NPOE), 59%	PVC	Udenafil /Udenafil-TmClPB	Slope of 60.3 mV/pC, Linear response range: 10-5.7 -10-2 M	(Han et al., 2010).

Table 2. Plasticizer used in some ion selective electrodes with polymeric membrane.

6. Conclusions

We presented here the importance and proper use of plasticizers in the preparation of membranes for ion-selective electrodes. The nature of the plasticizer influences key performance indicators of the ion-selective membrane electrodes such as slope, the domain of linear response and the selectivity. A plasticizer for the membrane preparation has to be compatible with the polymer and also with the electrodic component (ionophore), and have a high lipophilicity and low solubility in aqueous solution. The selection of the best plasticizer for the development of a polymeric membrane specific to a certain ion usually involves experimental tests in order to find the plasticizer and the ionophore with which the best response characteristics of the ion -selective electrodes are obtained. Several examples of selecting the best suited plasticizer for the design of selective electrodes sensitive to inorganic and organic ions have been presented in a detailed manner.

7. References

Baek, J.; Kim, J-S.; Paeng, I. K. & Paeng, K-J. (2008). The Composition Dependence Selectivity Changes by Plasticizer at the Cation Sensors Based on Tetracycline Antibiotics. *Bulletin of the Korean Chemical Society,* Vol. 29, No. 1, pp. 165-167

Egorov, V. V.; Astapovich, R. I.; Bolotin, A. A.; Vysotskii, D. L.; Nazarov, V. A.; Matulis, V. E. & Ivashkevich, O. A. (2010). The Influence of the Plasticizer Nature on the Selectivity of Ion-Selective Electrodes to Physiologically Active Amine Cations: Regularities and Abnormalities. *Journal of Analytical Chemistry,* Vol. 65, No. 4, pp. 404-413, ISSN 1061-9348

El-Nemmaa, EM., Badawib, BN., Saad S.M. Hassan, SSM. (2009). Cobalt phthalocyanine as a novel molecular recognition reagent for batch and flow injection potentiometric and spectrophotometric determination of anionic surfactants. *Talanta,* Vol. 78 pp. 723-729.

Evsevleeva, L. G.; Bykova, L. M. & Badenikov V. Ya. (2005). Aluminum-Selective Electrode. *Journal of Analytical Chemistry,* Vol. 60, No. 9, pp. 866-867

Han, W. S.; Lee Y. H.; Jung K. J.; Ly S. Y.; Hong T. K. & Kim M. H. (2008). Potassium Ion-Selective Polyaniline Solid-Contact Electrodes Based on 4′,4″(5″)-Di-tert - butyldibenzo-18-crown-6-ether Ionophore. *Journal of Analytical Chemistry,* Vol. 63, No. 10, pp. 987-993, ISSN 1061-9348

Han, W-S.; Kim, J-K.; Chung, K-C.; Hong, J-Y.; Hong, J-K.; Kim, J-H. & Hong, T-K. (2010). Poly(aniline) Solid Contact Ion Selective Electrode for Udenafil. *Journal of Analytical Chemistry,* Vol. 65, No. 10, pp. 1035-1040, ISSN 1061-9348

Kormosh, Zh. A.; Hunka, I. P. & Bazel, Y. R. (2009). An Ion-Selective Sensor for Assay of Diclofenac in Medicines. *Pharmaceutical Chemistry Journal,* Vol. 43, No. 7, pp. 54-56

Kulapina, E. G. & Barinova, O. Y. (1997). Structures of Chemical Compounds, Methods of Analysis and Process Control - Ion-Selective Electrodes in Drug Analysis. *Pharmaceutical Chemistry Journal,* Vol. 11, No. 12, pp. 40-45

Kulapina, E. G. & Barinova, O. V. (2001). Ion-Selective Electrodes for the Determination of Nitrogen-Containing Medicinal Substances. *Journal of Analytical Chemistry,* Vol. 56, No. 5, pp. 457-460

Kumar, K. G.; Augustine, P. & John, S. (2007), A Novel Potentiometric Sensor for the Determination of Nimesulide. *Portugaliae Electrochimica Acta,* No. 25, pp. 375-381

Lenik, J.; Wardak, C. & Marczewska, B. (2008). Propreties of Naxopen Ion-Selective Electrodes. *Central European Journal of Chemistr*, Vol. 6, No. 4, pp. 513-519

Lizondo-Sabater, J.; Martinez-Manez, R.; Sancenon, F.; Segui, J. & Soto, J. (2008). Ion-Selective Electrodes for Anionic Surfactants Using a Cyclam Derivative as Ionophore. *Talanta*, Vol. 75, pp. 317-325

Masadome, T.; Yang, J-G. & Toshihiko, I. (2004). Effect of Plasticizer on the Performance of the Surfactant-Selective Electrode Based on a Poly(Vinyl Chloride) Membrane with no Added Ion-Exchanger. *Mirochimia Acta*, No. 144, pp. 217-220

Matesic-Puac,. R., Sak-Bosnarb, M., Bilica, M. & Bozidar S. Grabaric, B. (2005). Potentiometric determination of anionic surfactants using a new ion-pair-based all-solid state surfactant sensitive electrode. *Sensors and Actuators B*, Vol 106 pp. 221–228.

Mihali, C. (2006), Researches Regarding Preparation of Electrochemical Sensors for Anionic Surfactants. *Ph. D. Thesis*, pp. 5-11

Mihali, C.; Oprea, G. & Cical, E. (2008). Determination of Critical Micelar Concentration of Anionic Surfactants Using Surfactants - Sensible Electrodes. *Chem. Bull. "POLITEHNICA" Univ. (Timisoara)*, Vol. 53(67), pp. 159-162

Mihali, C.; Oprea, G. & Cical, E. (2009). PVC matrix ionic – surfactant selective electrodes based on the ionic pair tetraalkyl-ammonium-laurylsulphate, Studia Universitatis Babes-Bolyai - Chemia series, Vol. LIV, No. 3, pp. 141-150

Moody, G. & Thomas J. (1986). Progress in designing calcium ion-selective electrodes. *Ion Selective Electrode Rev.*, Vol. 1, pp. 3-30

Nazarov, V. A.; Sokolova, E. I.; Androchik, K. A.; Egorov, V. V.; Belyaev, S. A. & Yurkshtovich, T. L. (2010). Ibuprofen – Selective Electrode on the Basis of a Neutral Carrier, N-Trifluoroacetylbenzoic Acid Heptyl Ester. *Journal of Analytical Chemistry*, Vol. 65, No. 9, pp. 960-963, ISSN 1061-9348

Oesch, U., Ammann, D. & Simon, W. (1886). Ion-Selective Membrane Electrodes for Clinical Use. *Clin. Chem.* Vol. 32, No. 8, pp. 1448-1459

Oprea, G., Mihali. C. & Hopirtean, E. (2007). Anionic Surfactants Selective Electrodes Based On Tricaprylmethyl Ammonium Laurylsulphate Ionophore. *Rev. Chim.*, Vol. 58, No. 3, pp. 335-338, ISSN 0034-7752

O'Rourke, M.; Duffy, N.; De Marco, R. & Potter, I. (2011). Electrochemical Impedance Spectroscopy—A Simple Method for the Characterization of Polymer Inclusion Membranes Containing Aliquat 336. *Membranes* 1, pp. 132-148; doi:10.3390/membranes102013, ISSN 2077-0375

Trebbe, U.; Niggermann, M.; Cammann, K.; Fiaccabrino, G. C.; Kundelka-Hep, M.; Dzyadevich, S. & Shulga, O. (2001). A New Calcium Sensor Based on Ion-Selective Conductometric Microsensor-Membranes and Features. *Fresenius Journal of Analytical Chemistry*, Vol. 371, pp.734-739

8

The Effect of Concentration and Type of Plasticizer on the Mechanical Properties of Cellulose Acetate Butyrate Organic-Inorganic Hybrids

Patrycja Wojciechowska
The Poznan University of Economics, Poznań
Poland

1. Introduction

Organic/inorganic polymer hybrids is a rapidly growing area of research because they offer opportunities to combine desirable properties of organic polymers (toughness, elasticity, formability) with those of inorganic solids (hardness, chemical resistance, strength). There are several routes to prepare hybrid materials, but one of the most common method is sol-gel technique generating inorganic phase within organic polymer matrix. The advantage of sol-gel technique is mild processing characteristics and the possibility of tailoring morphology of the growing inorganic phase and thus properties of the material by the subtle control of various reaction conditions. This process includes hydrolysis of the precursor (metal alkoxide) followed by condensation reactions of the resulting hydroxyl groups. Considering the nature of the interface between the organic and inorganic phases, hybrid materials can be categorized into two different classes. The first class corresponds to non-covalently bound networks of inorganic and organic phases. These hybrids show weak interactions between the polymer matrix and inorganic phase, such as van der Waals, hydrogen bonding or weak electrostatic interactions and can be prepared by physical mixing of an organic polymer with a metal alkoxide. In the second class organic and inorganic phases are linked through strong chemical bonds (covalent or ionic). Chemical bonding can be achieved by the incorporation of silane coupling groups into organic polymers [1-3].

Cellulose has received a great deal of attention in recent decades as a substitute for petrochemical based polymers. Natural polymer shows however some limitations, for instance with regard to poor processability or high water absorbency. Cellulose esters such as cellulose acetate (CA), cellulose acetate propionate (CAP) and cellulose acetate butyrate (CAB) are less hydrophilic than cellulose, thermoplastic materials [4]. To improve their processability and mechanical properties, the addition of plasticizers is usable. Plasticizers as polymer additives serve to decrease the intermolecular forces between the polymer chains, resulting in a softened and flexible polymeric matrix. They increase the polymer's elongation and enhance processability by lowering the melting and softening points and viscosity of the melts [5].

Plasticizers are often inert organic compounds with low molecular weight, high boiling points and low vapor pressures that are used as polymer additives. The main role of the plasticizer is to improve mechanical properties of the polymers by increasing flexibility, decreasing tensile strength and lowering the second order transition temperature [6]. The International Union of Pure and Applied Chemistry (IUPAC) developed a definition for a plasticizer as a "substance or material incorporated in a material (usually a plastic or an elastomer) to increase its flexibility, workability, or distensibility" [7]. Attributes of a good plasticizer are good compatibility with polymer, which depends on polarity, solubility, structural configuration and molecular weight of plasticizer and results from a similar chemical structure of polymer and plasticizer. Other important factor is plasticizer permanence related to its resistance to migration. Therefore, a good plasticizer should have high boiling point and low volatility (low vapor pressure) to prevent or reduce its loss during processing. Plasticizers should also be aroma free and non-toxic. Another important feature is low rate of migration out of material to preserve desirable properties of plasticized polymer and avoid contamination of the materials from the point of potential health and environmental impacts in contact with it. The permanence of plasticizer in polymer is dependent on the size of the plasticizer molecule, thus the larger molecules, the greater permanence of the plasticizer. The higher diffusion rate of plasticizer in the polymer, the lower permanence due to the migration out of the polymer matrix [8, 9]. Plasticizers influence also processing of the polymers by changing various parameters: viscosity, filler incorporation, dispersion rate, flow, power demand and heat generation [7]. A good plasticizer should also be insensitive to solar UV radiation, stable in a wide temperature range and inexpensive [6]. The efficiency of a plasticizer is defined as the quantity of plasticizer required to provide desired mechanical properties of obtained material [8]. Taking into consideration that effective plasticization is depended on such factors as: chemical structure of the plasticizer, its compatibility and miscibility with the polymer, molecular weight and concentration of plasticizer, rate of diffusion of the plasticizer into the polymer matrix, different polymers require different plasticizers [8].

2. Plasticizer classification

There are two techniques for plasticization: external and internal. External plasticization is a method that provides plasticity through physical mixing. Thus, external plasticizers are not chemically bound to the polymer and can evaporate, migrate or exude from polymer products by liquid extraction [6]. Plasticization of polymers by incorporation of comonomers or reaction with the polymer, providing flexible chain units is called an internal plasticization. Internal plasticizers are groups (flexible segments) constituting a part of a basic polymer chain, which may be incorporated regularly or irregularly between inflexible monomers (hard segments) or grafted as side chains thus reducing intermolecular forces [7, 10-12]. According to the compatibility with the polymer, external plasticizers can be classified into two principal groups: primary and secondary ones, called also extenders. Primary plasticizers have a sufficient level of compatibility with polymer to be able to be used as sole plasticizer in all reasonable proportions, giving a desirable modifying effect. They interact directly with chains. Secondary plasticizers have limited compatibility and will exude from the polymer if used alone. They are used along with the primary plasticizer, as a part of plasticizer system, to meet a secondary performance requirements (cost, low-temperature properties, permanence). Extenders can be used as lower cost, partial replacement for a primary plasticizer. It is possible that a plasticizer used in one formulation as a primary plasticizer could be used in a second formulation as a second one [10, 11]. Plasticizers, especially used in biopolymer-based films, can also be classified as water soluble and water insoluble. Hydrophilic plasticizers dissolve in polymeric aqueous dispersions and may cause an increase of water diffusion in the polymer when added in high concentration. On the contrary, hydrophobic plasticizers can lead to a decrease in water uptake, due to the closing of micro-voids in the polymer [7].

3. Mechanisms of plasticization

There are several theories that describe the effects of plasticizers and a combination of them allows to explain the concept of polymer plasticization [8, 10, 13-15]:

a. Lubricity theory, developed by Kilpatrick, Clark and Houwink, among others, states that plasticizer acts as a lubricant, reducing intermolecular friction between polymer molecules responsible for rigidity of the polymer. On heating, the plasticizer molecules slip between polymer chains and weaken the polymer-polymer interactions (van der Waals' forces), shielding polymer chains from each other. This prevents the re-formation of a rigid network, resulting in more flexible, softener and distensible polymer matrix.
b. Gel theory, developed by Aiken and others, holds that polymers are formed by an internal three-dimensional network of weak secondary bonding forces (van der Waals' forces, hydrogen bonding) sustained by loose attachments between the polymer molecules along their chains. These bounding forces, are easily overcome by external strain applied to the material, allowing the plasticized polymer to be bend, stretch, or compress. Plasticizer molecules attach along the polymer chains, reducing the number of the polymer-polymer attachments and hindering the forces holding polymer chains together. The plasticizer by its presence separates the polymer chains and increases the space between polymer molecules, thus reducing the rigidity of the gel structure. Moreover, plasticizer molecules that are not attached to polymer tend to aggregate allowing the polymer molecules to move more freely, thus enhancing the gel flexibility.
c. Free volume theory holds that the presence of a plasticizer lowers the glass transition temperature (Tg) of the polymer. Free volume is a measure of internal space available within a polymer matrix. There are three main sources of free volume in polymer: motion of polymer end groups, motion of polymer side groups, and internal polymer motions. When the free volume increases, more space or free volume is provided for molecular or polymer chain movement. A polymer in the glassy state has an internal structure with molecules packed closely and small free volume. This makes the material rigid and hard. When the polymer is heated to above the glass transition temperature, the thermal energy and molecular vibrations create additional free volume which allows greater internal chain rotation and an increase in the segment mobility. This makes the system more flexible and rubbery. When small molecules such as plasticizers are added, the free volume available to polymer chain segments increases and therefore the glass transition temperature lowers.

d. Mechanistic theory of plasticization considers that plasticizer molecules are not bound permanently to the polymer, but rather there is a dynamic exchange process whereby, a constant associations and disassociations of polymer-polymer, polymer-plasticizer and plasticizer-plasticizer molecules form. Some plasticizers form stronger associations with polymer than others. At low plasticizer levels, the plasticizer-polymer interactions are the dominant interactions, what explains "antiplasticization". At high plasticizer loadings plasticizer-plasticizer associations predominate.

Plasticizers have been used as a polymer additives since 1800s [7]. The worldwide plasticizer demand in 2009 was about 5.7 million tons constituting 51.8% share of global polymer additives market [16]. About 100 plasticizers among 1200 different plasticizers produced worldwide are classified as commercially important [7]. Approximately 90% of all plasticizers are used in plasticized or flexible poly(vinyl chloride) (PVC) products [13, 16]. Plasticizers are also required in such polymer systems as poly(vinyl butyral), poly(vinyl acetate), acrylic polymers, poly(vinyldiene chloride), nylon, polyamides, cellulose molding compounds, polyolefins and certain fluoroplastics [7, 17]. The most significant and the largest group of PVC plasticizers is esters of phthalic acid with the share of 97% of all plasticizers used. Phthalate esters plasticizers are mostly based on carboxylic acid esters containing linear or branched aliphatic alcohols of chain lengths C6-C11. Phthalate esters have been used as plasticizers in plastic materials since the 1920s. Widely used phthalates are: di(2-ethylhexyl)phthalate (DEHP), also known as dioctyl phthalate (DOP), di-isononyl phthalate (DINP), di-isodecyl phthalate (DIDP), di-butyl phthalate (DBP) and butyl benzyl phthalate (BBP). The most broadly used since 1930s phthalate plasticizer has been DOP [6, 7]. Phthalate esters are usually added in concentrations up to 50% of the final weight of the products [18, 19]. According to Ceresana Research report, plasticizer market in 2010 was dominated by phthalate esters, with 54% share of DOP, as the most widely used. Ceresana Research forecasts that over the next years DOP will be increasingly replaced by alternative plasticizers due to worldwide growing concerns about the potential toxicity of phthalate esters to humans and the environment [20]. The application of phthalate plasticizers is being questioned because as low molecular weight compounds they migrate out of the polymer matrix. Since they are commonly used in a variety of products: flexible plastics, toys, flooring and car dashboards, food contact materials, packaging systems, synthetic leather, medical devices like blood transfusion bags and haemodialysis tubing, cosmetics, as a result, they have been found in terrestrial and aquatic ecosystems, in domestic foods and wastes, and also in animals and humans. Main human exposure pathways to phthalates include inhalation of air contaminated due to off-gassing from plasticized products, also food and drinking water containing plasticizers that exude from packaging materials designed for victuals or are extracted by the foodstuff [6, 18, 21]. Unfortunately, the exposure to a number of phthalates among the general population is wide, with the highest doses for infants and children, due to additional intake caused by the mouthing behavior of toys. Important exposure pathways of phthalates are food and intensive medical care [6, 22]. There are numerous reports showing that phthalates exert adverse effects on animals' liver, heart, kidney, lungs [23]. A number of studies have been also conducted to evaluate the potential toxicity of phthalate plasticizers on human health. The results showed several implications: hormonal disorders, inducing hepatic peroxisome proliferation, reproductive toxicity, carcinogenicity, allergic symptoms in children [6, 21, 22, 24, 25]. Public health concerns implied changes in legal provisions. Since 1999, the use of six phthalate plasticizers: DINP, DIDP, DEHP, DBP, BBP and DnOP (di-n-octyl phthalate) in childcare products and toys that can be placed in the mouth of children under the age of three in European Union is restricted. Further regulations in 2005 introduced directive that forbids the use of DEHP, DBP and BBP in any toys and childcare articles within European Union. DEHP, DBP and BBP are also forbidden to be used in cosmetic products and restricted in preparations such as paints and varnishes for end-consumers [18, 22]. The above mentioned reasons caused growing interest in less questioned substitutes of phthalate esters. Commercial used phthalates can be replaced by nontoxic alkyl esters of adipic and citric acids or natural-based plasticizers like epoxidized triglyceride vegetable oils from soybean oil, linseed oil, castor-oil, sunflower oil, and fatty acid esters [7]. The advantages of these alternative additives are good technical performance, processing ease and low toxicity. An important feature of alternative plasticizers is also biodegradability, due to the growing interest of materials obtained from degradable polymers and biopolymers from renewable resources [26, 27]. Other substitutes to phthalates are polymeric plasticizers (for example based on phthalic acid) and oligomers that exhibit low volatility and thus show low rate of migration out of the polymer and leaching tendency. Promising properties show also phenol alkyl sulfonate plasticizers which exhibit excellent gelling capacity thus reducing processing time and temperature. This class of additives shows also reduced leaching tendency and are predestined for medical applications such as polymeric materials exposed to warm, aqueous media for an extended period of time. An interesting, environmentally friendly alternative to phthalates (especially for PVC and poly(methyl methacrylate) are also ionic liquids, however they are still under research [6]. Among esters of bioderived citric acid tributyl citrate, acetyl tributyl citrate, triethyl citrate, acetyl triethyl citrate, and tri(2-ethylhexyl) citrate are of importance. Citric acid esters have been approved as plasticizers for medical plastics, personal care, and according to the U.S. Food and Drug Administration, as additives in food [9, 28]. Citrate esters have been used as effective plasticizers for environmental friendly polymers such as poly(lactic acid), cellulose acetate. However, besides enhanced processability, accelerated degradation rates were also observed [29]. Another class of plasticizers applied in biodegradable polymers are polyols. Among them glycerol, ethylene glycol (EG), propylene glycol (PG), diethylene glycol (DEG), triethylene glycol (TEG), tetraethylene glycol and polyethylene glycol (PEG) are the most often used as polymer additives [6, 7]. Glycerol, which have found application as effective plasticizer for starch or gelatin, and TEG are suitable for use in the food industry as they are on the FDA's Generally Regarded As Safe (GRAS) list [6].

In spite of a wide range of new plasticizers available for polymer industry it must be emphasized that alternative additives may replace traditional ones only in some specific applications due to the several requirements: compatibility, solvation, permanence and price.

There are numerous reports in the literature associated with polymer blends based on cellulose derivatives plasticized with conventional and alternative plasticizers: cellulose acetate plasticized with DEP, triethyl citrate (TEC), and poly(caprolactone triol) (PCL-T), cellulose acetate butyrate plasticized with TEC [27, 30-33].

In our previous work we examined the effect of inorganic phase amount and diethyl phthalate and citrate plasticizer on the degradability of organic-inorganic cellulose acetate butyrate films in sea water [34]. The results of our study showed that the higher the amount of silica incorporated into the CAB with the DEP plasticizer, the higher degradability of the samples. The experiment also showed a synergistic effect of the applied plasticizer on the degradation rate of the CAB/silica hybrids. The CAB/silica hybrids with diethyl phthalate were degraded faster than the hybrids with tributyl citrate due to the higher brittleness of those samples. The aim of the present study is to examine the effects of six different plasticizers: citrate esters and phthalates, on the mechanical properties of cellulose acetate butyrate hybrids.

4. Materials and methods

Cellulose acetate butyrate (CAB, Mn≈ 70000), TEOS (98%) and TEA (99%) were purchased from Sigma-Aldrich. TEC (98%) and TBC (97%) were purchased from Fluka. DEP (99%), DBP (99%) and DOP (99%) were purchased from POCH and used as received. Organic-inorganic hybrids were synthesized according to the procedure we described in patent number 209829 [35]. Cellulose acetate butyrate hybrids were prepared with various amounts of TEOS: 6,25 wt.% and 12,5 wt.%, and various amounts of the chosen plasticizer (25-35%), such as biodegradable citrates: TEC, TEA, TBC and conventional phthalates: DEP, DBP, DOP. Obtained films showed thickness in the range of 0,15-0,18 mm. Samples prepared with concentration below 25% of all investigated plasticizers were too brittle for tensile testing.

Sample compositions and codes are as follows:

a. samples prepared from plasticized CAB: short name of plasticizer (TEC, TEA, TBC, DEP, DBP or DOP)/plasticizer content, e.g. TEC25, DOP35,
b. organic-inorganic hybrids prepared from composition of plasticized CAB and TEOS in 87.5/12.5 polymer/TEOS ratio: amount of TEOS/short name of plasticizer (TEC, TEA, TBC, DEP, DBP or DOP)/plasticizer content, e.g. 12.5TEC25, 12.5DBP30,
c. organic-inorganic hybrids prepared from composition of plasticized CAB and TEOS in 93.75/6.25 polymer/TEOS ratio: amount of TEOS/short name of plasticizer (TEC, TEA, TBC, DEP, DBP or DOP)/plasticizer content, e.g. 6.25TEC25, 6.25DBP30.

A typical preparation of organic-inorganic hybrid was as follows [36]: polymer was placed in a polyethylene beaker and dissolved in acetone. Plasticizer and TEOS was then added and mixed vigorously. To this solution catalytic amount of HCl (0.1 M) was added to initiate the sol-gel process and mixed until it appeared clear and homogenous. The solution was cast in an evaporating PTFE dish and left exposed to atmospheric conditions followed by drying in a vacuum drier at 40°C for 12 hours to ensure complete solvent evaporation.

Mechanical properties were investigated using a universal tensile machine (Instron 5565) at a crosshead speed of 100 mm/min at room temperature (according to the test method described in International Standards PN-EN ISO 527-1:1998, PN-EN ISO-3:1998). Sample dimensions: length 150 mm, width 10 mm. At least five tests were performed for each type of the sample, to ensure the reliability of the test results, and the average was used.

The properties of the materials used in this study are showed in Table 1.

Full name	Short name	Chemical structure	Molecular weight	Vapor pressure	Boiling point
Cellulose acetate butyrate	CAB	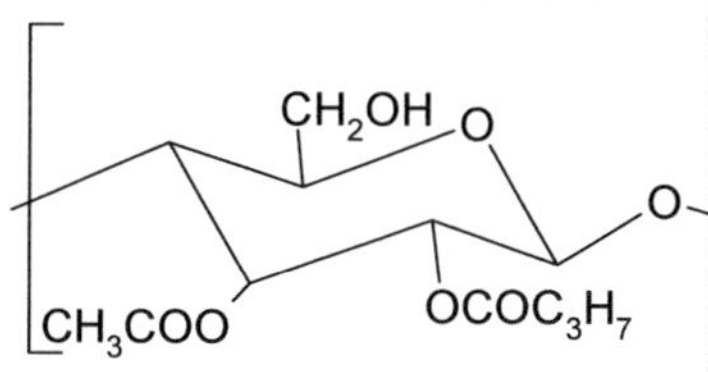	average M_n ~70,000	-	- (melting range 150-160°C)
Tetraethoxysilane	TEOS	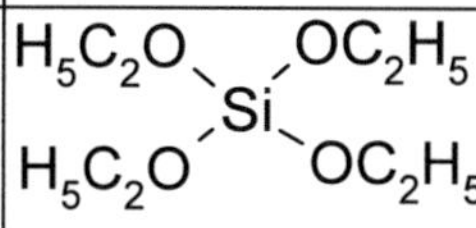	208.33	<1 mmHg (20°C)	168°C
Triethyl citrate	TEC	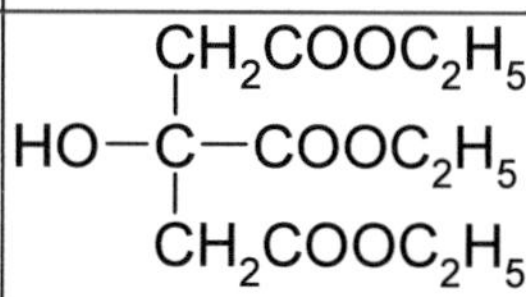	276.28	1 mmHg (107°C)	235°C/ 150 mmHg
Acetyl triethyl citrate	TEA	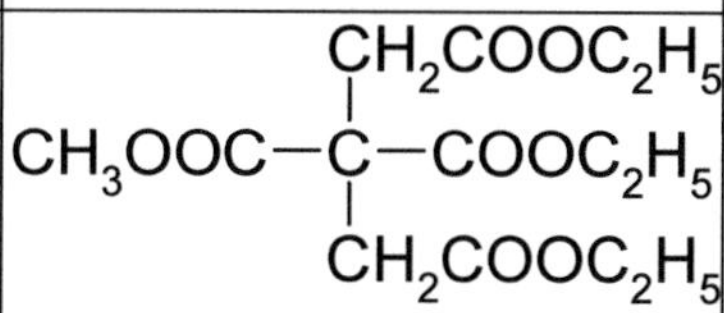	318.32	not available	228-229°C/ 100 mmHg
Tributyl citrate	TBC	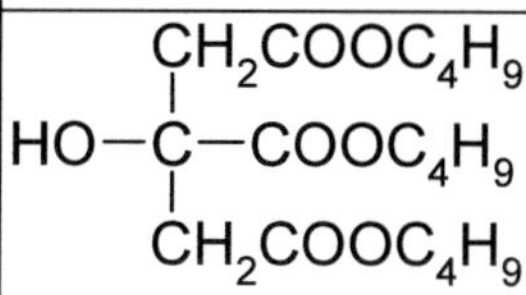	360.44	not available	234°C / 17 mmHg
Diethyl phthalate	DEP	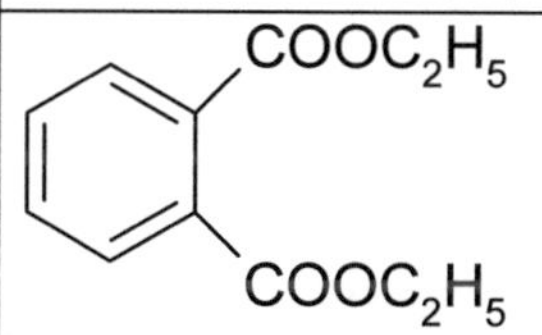	222.24	1 mmHg (100°C)	298-299°C
Dibutyl phthalate	DBP	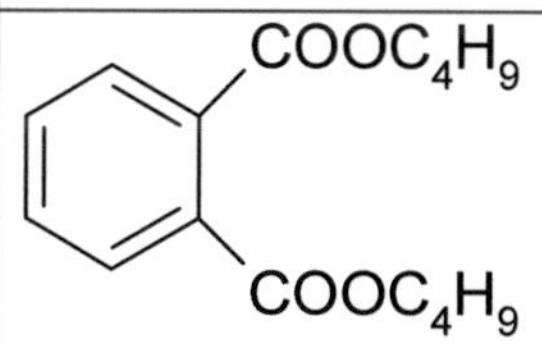	278.34	1 mmHg (147°C)	340°C
Dioctyl phthalate	DOP	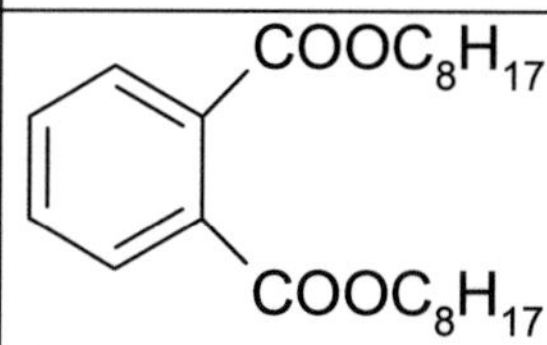	390.56	1.2 mmHg (93°C)	384°C

Table 1. Properties of the materials used in this study.

5. Results and discussion

Comparison of mechanical properties of organic-inorganic hybrids and cellulose acetate butyrate with different plasticizers is shown in Table 2 and Figures 1-18.

Type of the plasticizer	Polymer/TEOS ratio					
	87.5/12.5		93.75/6.25		100	
	Tensile strength (MPa)	Elongation at break (%)	Tensile strength (MPa)	Elongation at break (%)	Tensile strength (MPa)	Elongation at break (%)
TEA 25	24.9 ± 0.5	7.0 ± 0.4	23.0 ±1.3	16.2 ± 1.1	20.1 ± 1.5	26.3 ± 3.0
TEA 30	24.9 ± 1.1	37.8 ± 8.5	23.1 ± 1.3	44.8 ± 9.1	21.6 ± 1.0	34.3 ± 6.1
TEA 35	21.2 ± 1.4	45.6 ± 5.4	21.8 ± 0.8	53.6 ± 4.1	20.4 ± 0.7	48.7 ± 1.1
TBC 25	17.4 ± 0.7	16.7 ± 4.1	16.4 ± 0.4	13.9 ± 1.4	15.5 ± 1.7	24.3 ± 2.3
TBC 30	25.3 ± 2.8	40.9±13.6	23.6 ± 2.5	42.4 ± 7.6	21.8 ± 2.3	30.2 ± 6.9
TBC 35	15.7 ± 1.5	53.1 ± 8.4	14.3 ± 0.8	53.9 ± 2.1	13.4 ± 1.0	38.1 ± 4.4
TEC 25	24.0 ± 0.8	5.0 ± 0.6	17.0 ± 1.6	7.6 ± 1.1	21.5 ± 0.8	14.7 ± 2.4
TEC 30	14.4 ± 0.1	25.8 ± 1.6	13.3 ± 0.4	21.6 ± 3.5	12.7 ± 0.5	29.0 ± 6.3
TEC 35	12.8 ± 0.3	29.3 ± 4.2	11.5 ± 0.3	25.8 ± 0.8	11.2 ± 0.4	32.0 ± 3.8
DEP 25	15.9 ± 0.5	5.5 ± 0.9	13.8 ± 1.7	5.5 ± 0.3	12.5 ± 1.5	8.7 ± 1.7
DEP 30	20.9 ± 0.8	16.3 ± 1.5	19.2 ± 1.1	14.0 ± 2.0	17.6 ± 1.0	10.6 ± 1.7
DEP 35	21.7 ± 1.1	19.9 ± 0.4	17.4 ± 0.9	15.0 ± 3.0	14.1 ± 0.9	17.8 ± 0.6
DBP 25	23.2 ± 1.7	19.3 ± 3.5	20.5 ± 1.6	23.6 ± 2.8	21.0 ± 1.7	25.5 ± 2.5
DBP 30	26.4 ± 0.8	33.9 ± 5.0	25.0 ± 0.6	35.6 ± 4.6	24.8 ± 0.3	31.8 ± 5.5
DBP 35	20.3 ± 0.7	48.6 ± 5.6	16.4 ± 0.5	36.1 ± 3.1	14.6 ± 0.2	37.2 ± 2.8
DOP 25	27.3 ± 2.8	28.7 ± 0.4	23.7 ± 1.6	21.4 ± 1.7	22.3 ± 1.6	13.8 ± 2.8
DOP 30	31.1 ± 1.2	52.1 ± 1.5	28.1 ± 2.5	42.5 ± 4.5	28.3 ± 1.8	34.3 ± 2.0
DOP 35	23.5 ± 3.9	50.1 ± 3.3	19.9 ± 2.1	40.4 ± 6.1	16.7 ± 1.1	38.0 ± 5.0

Table 2. Mechanical properties of CAB samples containing various plasticizers.

The aim of adding plasticizer to CAB-hybrids is to reduce natural brittleness of the polymer and to enhance plastic elongation, while providing optimal tensile strength and stiffness.

The plasticizing efficiency of the investigated phthalates and citrates evaluated by tensile testing is summarized in Table 2. At concentration 25% samples of the cellulose acetate butyrate plasticized with TEA, TEC, DBP and DOP exhibited similar tensile strength in the range of 20 - 22 MPa, however high values of elongation at break (24 - 26%) showed only samples containing TBC, DBP and TEA. In case of CAB hybrids the introduction of inorganic phase into polymer matrix caused hardening and reinforcing of the material, thus an increase of tensile strength in comparison with unmodified CAB was observed. Regarding organic-inorganic hybrids prepared from 93.75/6.25 and 87.5/12.5 polymer/TEOS formulations the highest values of tensile strength (23 - 24 MPa and 25 - 27 MPa) were obtained for samples 6.25TEA25, 6.25DOP25, and 12.5TEA25, 12.5DOP25, respectively. However, at the same time, obtained samples exhibited lower values of elongation at break as compared with plasticized CAB, due to the higher brittleness of the material. The results showed that the presence of 25% of plasticizer in organic-inorganic CAB hybrids was insufficient for providing acceptable flexibility.

Considering the effect of plasticizer concentration it can be concluded that all of the plasticizers investigated, excluding TEC, caused an antiplasticization at concentration 30% of the plasticizer, resulting in an increase in tensile strength in comparison with the values at 25%. To the contrary, samples plasticized with TEC showed a common trend: with increasing plasticizer content, the tensile strength decreased, while elongation at break increased. Antiplasticizing effects were previously observed by Donempudi et al. for PVC membranes plasticized with phthalates [37], reported for citrate esters used as plasticizers for poly(methyl methacrylate) (PMMA) [38], and also has been found for polycarbonate, polysulfone, polystyrene plasticized with various plasticizers [39]. Even though the phenomenon of antiplasticization has been already long observed in synthetic polymers, the mechanisms involved are not perfectly known. According to Anderson et al. the phenomenon can be attributed to a chain end effect. Antiplasticizers initially fill unoccupied lower volume at the chain end and then the overall polymer free volume. Chain end mobility is restricted, resulting, thus, in higher modulus and resistance, generally followed by polymer hardness. Jackson and Caldwell suggested that antiplasticization can be attributed to a free volume reduction due to antiplasticizers [40]. Another explanation is an increase in the degree of order or the crystallinity of the system, resulting in an increase in tensile strength. Antiplasticization of the samples may be attributed to the hindered local mobility of the macromolecules, and thus reduced flexibility, due to the strong interaction between polymer and plasticizer (i.e. hydrogen bonding, van der Waals' forces) [39, 41]. Antiplasticization in polymers depends on molecular weight and concentration of the diluent and occurs over a concentration range below the plasticization threshold. This point, dividing antiplasticization and plasticization behavior, is typical for each polymer-plasticizer system [42]. Gutierrez-Villarreal [38] reported an antiplasticization effect for PMMA plasticized with TEC at low concentration of plasticizer (about 13 wt%). The plasticization threshold for TEC plasticized samples based on CAB was not observed in the range of concentrations used in this study. For the samples prepared with lower concentration of TEC (below 25%) the measurement using a universal tensile machine was difficult to perform due to the high brittleness of the organic-inorganic hybrids (cutting of the samples might induce micro-cracking on the edge of the samples and influence the reliability of the test results).

Considering the fact that different factors may be involved in the antiplasticization phenomenon, the present study was not designed to provide evidence in support of any one of these mechanisms. Further experiments including dynamic mechanical analysis (DMA), differential scanning calorimetry (DSC) or X-ray measurements could confirm suggested hypothesis.

At concentration 30% the CAB samples plasticized with DOP and DBP showed the highest tensile strength (28.3 MPa and 24.8 MPa, respectively). Among citrate plasticizers the higher tensile strength values were obtained for CAB samples plasticized with TBC and TEA (21.8 MPa and 21.6 MPa, respectively). The lowest values of tensile strength showed CAB samples plasticized with TEC (12.7 MPa) and DEP (17.6 MPa) due the high brittleness of the material, indicating low plasticizing efficiency of those plasticizers. Interestingly, organic-inorganic hybrids showed both high values of tensile strength, regardless of the plasticizer type and concentration, as well as elongation at break in comparison with plasticized CAB. Organic-inorganic hybrid prepared from 87.5/12.5 polymer/TEOS formulation and DOP (12.5DOP30) exhibited the highest tensile strength (31.1 MPa) as well as very high elongation at break (52.1%). Regarding the citrate plasticizers at 30% concentration the best mechanical properties were obtained for TBC and TEA. In this case, organic-inorganic hybrids prepared from 87.5/12.5 polymer/TEOS formulation plasticized with TBC and TEA showed similar values of tensile strength and elongation at break: 25.3 MPa and 40.9%, and 24.9 MPa and 37.8% , respectively.

At higher concentration of plasticizers used in this study (35%) the additives caused plasticization reflected as a decreases in tensile strength and an increase in elongation at break values. Regarding CAB samples, the highest values of elongation at break showed material plasticized with TEA (48.7%). Among phthalates, at level of 35%, the highest value of elongation at break CAB reaches for DOP and DBP (38.4% and 37.2%, respectively). The highest values of elongation at break for the organic-inorganic hybrids obtained from 93.75/6.25 polymer/TEOS formulation were observed for samples plasticized with TBC, TEA and DOP (53.9%, 53.6% and 40.4%, respectively). In case of organic-inorganic hybrids obtained from 87.5/12.5 polymer/TEOS formulation the highest values of elongation at break provided TBC, DOP and DBP plasticizers (53.1%, 50.1% and 48.6%, respectively).

If one considers the effect of plasticizer molecular weight on the mechanical properties of investigated samples, one might conclude that the higher molecular weight, the better efficiency of the plasticizer. Regarding phthalate esters, plasticizer with the lowest molecular weight produced the less flexible samples and the efficiency varied in the order DEP>DBP>DOP. Similar behavior was previously observed for phthalate esters used as plasticizers for PVC membranes [37]. Donempudi at al. found that the tensile strength of the membranes decreased as the size of the alkyl group of the phthalate molecule increased from methyl to octyl, meanwhile the elongation at break values increased. They referred that an increase in the size of the alkyl chain length of the phthalate molecule brought about an increased dilution of the polymer solution. Hence, the high molecular weight implied a further reduction in the number of macromolecules per unit volume. Therefore, the use of higher concentration of larger size phthalate molecules in the PVC matrix caused significant dilution effect, and as a result an increase in the flexibility of the polymer [37]. Similar results were obtained also for citrate plasticizers applied in the study. The lowest plasticizing efficiency of TEC, among citrate plasticizers used in this work, may be attributed to its low molecular weight. On the contrary, the highest molecular weight TBC, containing longer alkyl groups was found to be the most efficient.

The stress-strain curves for the samples prepared with different plasticizers are presented in Fig. 1-18. The characteristic type of the curve for hard and rigid materials, exhibiting low values of elongation at break, showed organic-inorganic hybrids prepared with 25% of TEC and DEP (Figure 7, 10). Hard, tough behaviour is observed for the samples exhibiting sufficient and good plasticizing efficiency (Fig. 1-6, 8, 9, 11-18). All the curves showed cold drawing and strain hardening in the final section of the curve. However, for the samples prepared from the formulations exhibiting the best mechanical properties, the curves showed better defined yielding point. In case of organic-inorganic hybrids with the highest content of inorganic phase the curves exhibited elastic deformation in smaller strain ranges than for the plasticized CAB.

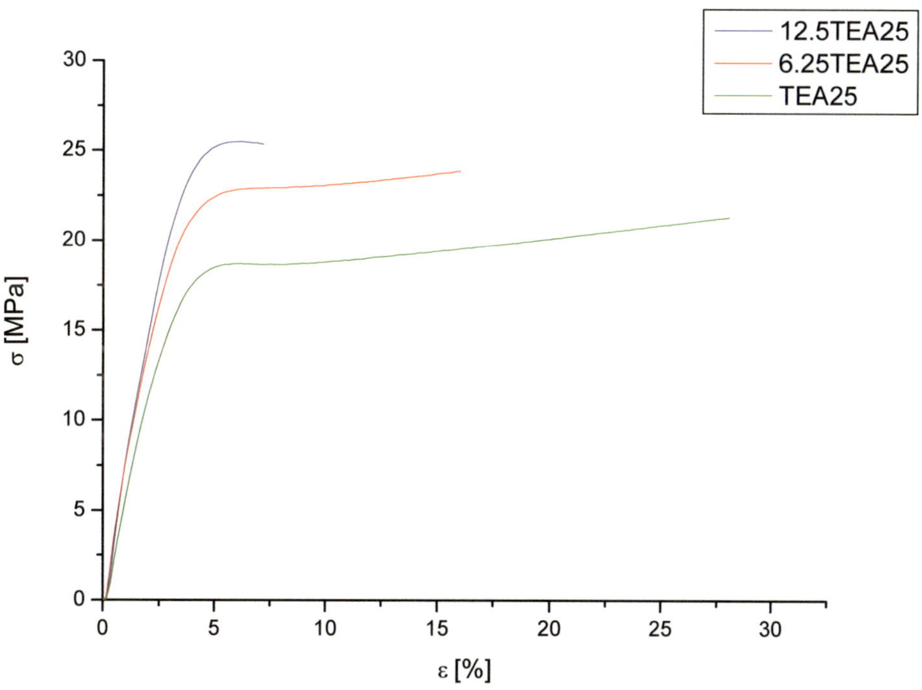

Fig. 1. The tensile stress-strain curves for samples prepared with 25% of TEA.

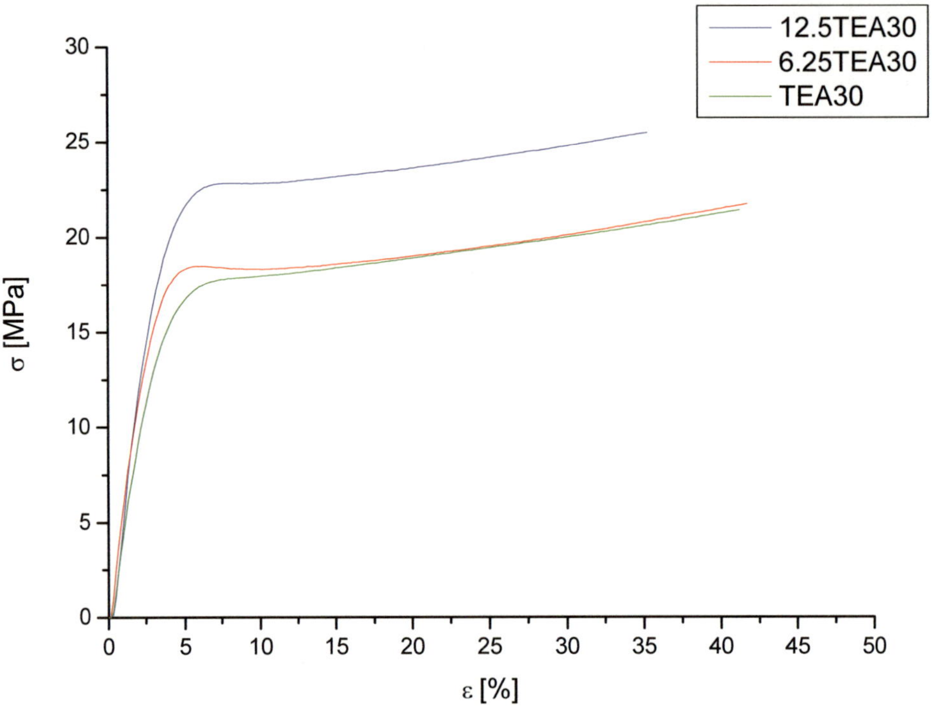

Fig. 2. The tensile stress-strain curves for samples prepared with 30% of TEA.

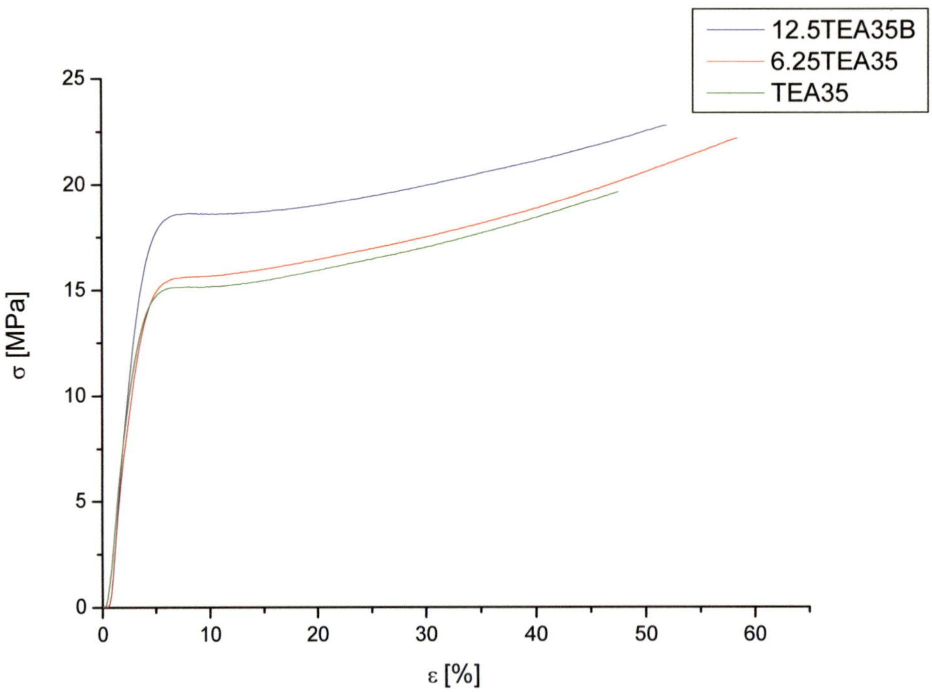

Fig. 3. The tensile stress-strain curves for samples prepared with 35% of TEA.

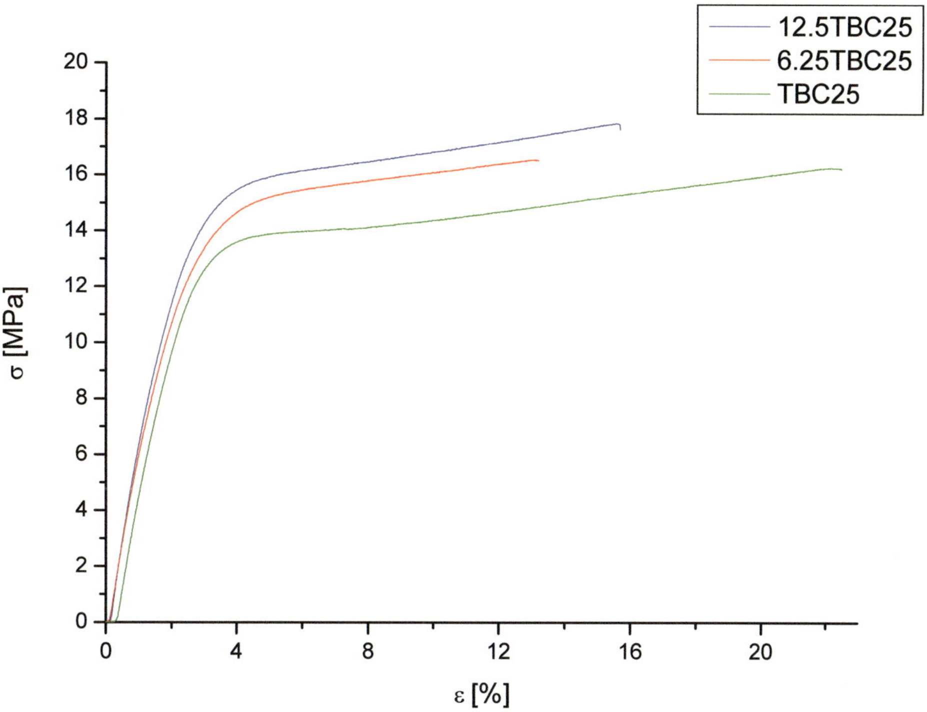

Fig. 4. The tensile stress-strain curves for samples prepared with 25% of TBC.

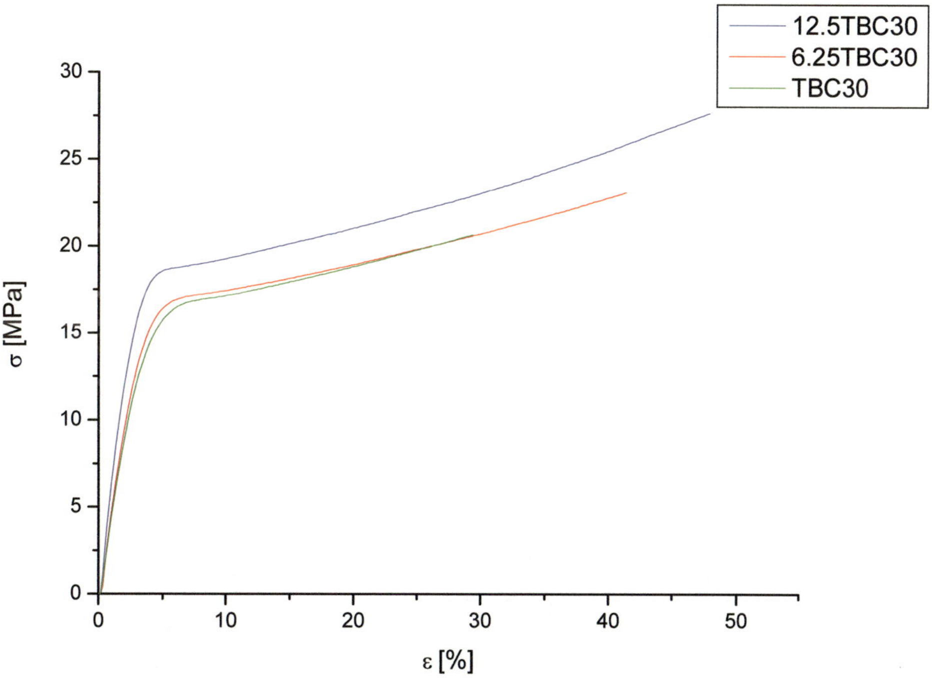

Fig. 5. The tensile stress-strain curves for samples prepared with 30% of TBC.

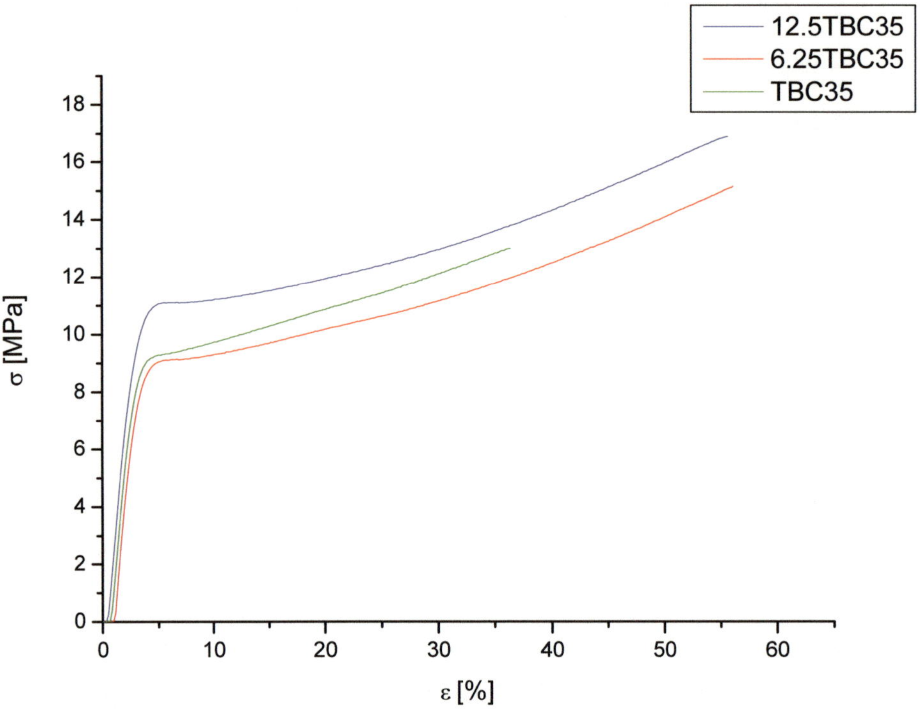

Fig. 6. The tensile stress-strain curves for samples prepared with 35% of TBC.

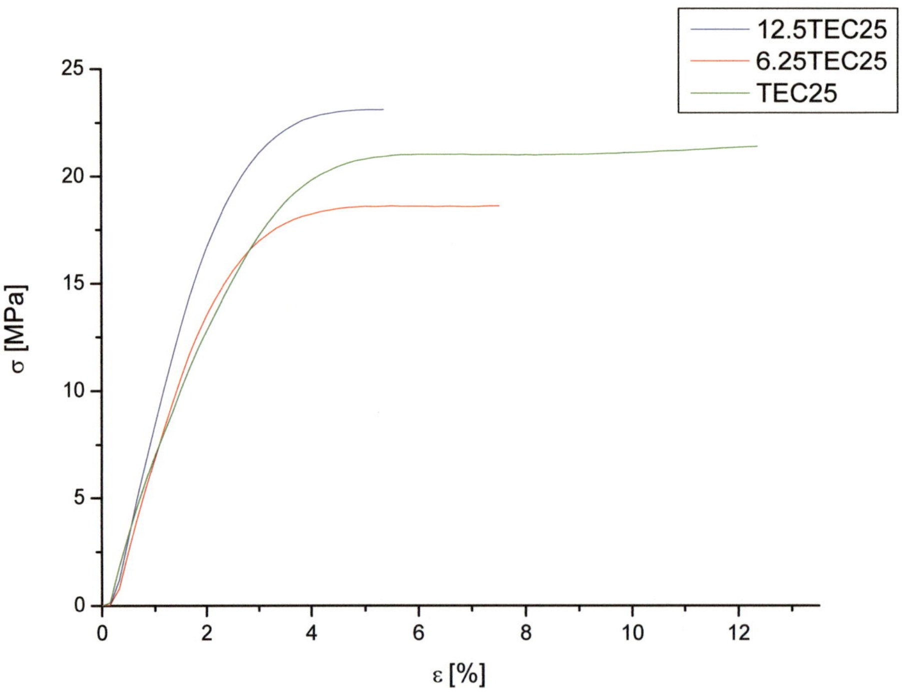

Fig. 7. The tensile stress-strain curves for samples prepared with 25% of TEC.

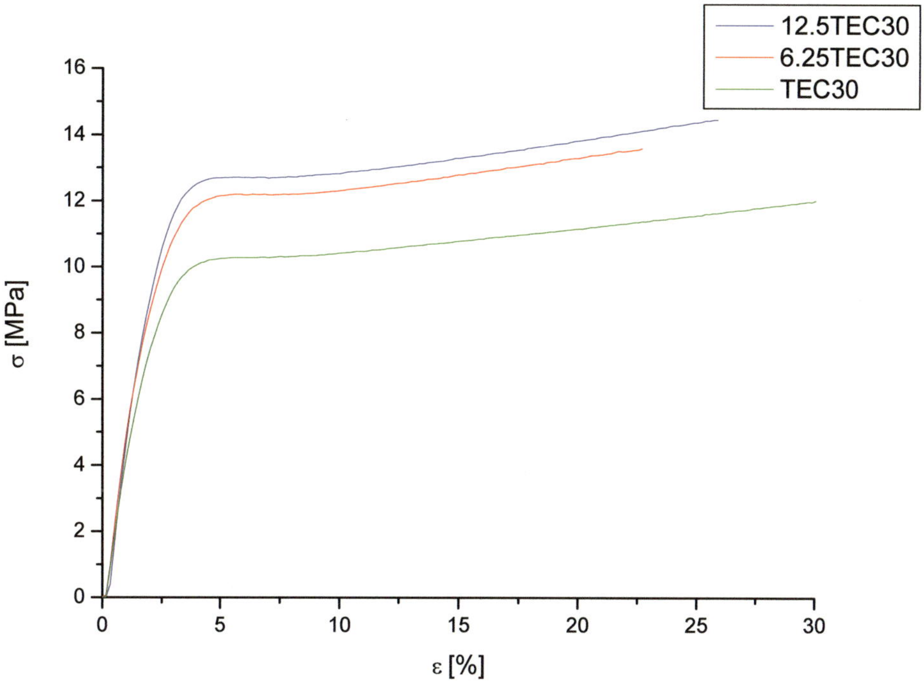

Fig. 8. The tensile stress-strain curves for samples prepared with 30% of TEC.

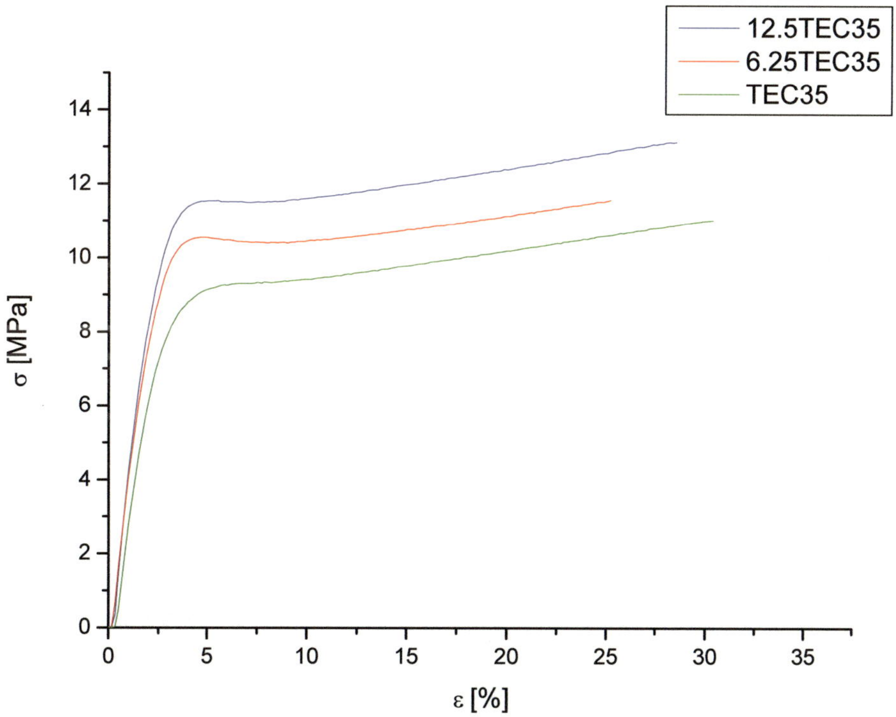

Fig. 9. The tensile stress-strain curves for samples prepared with 35% of TEC.

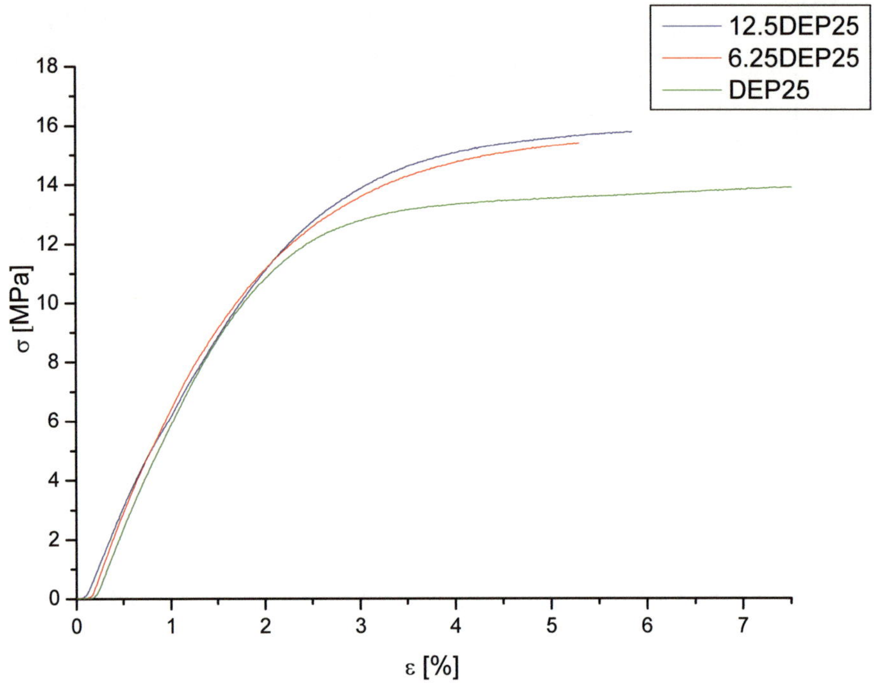

Fig. 10. The tensile stress-strain curves for samples prepared with 25% of DEP.

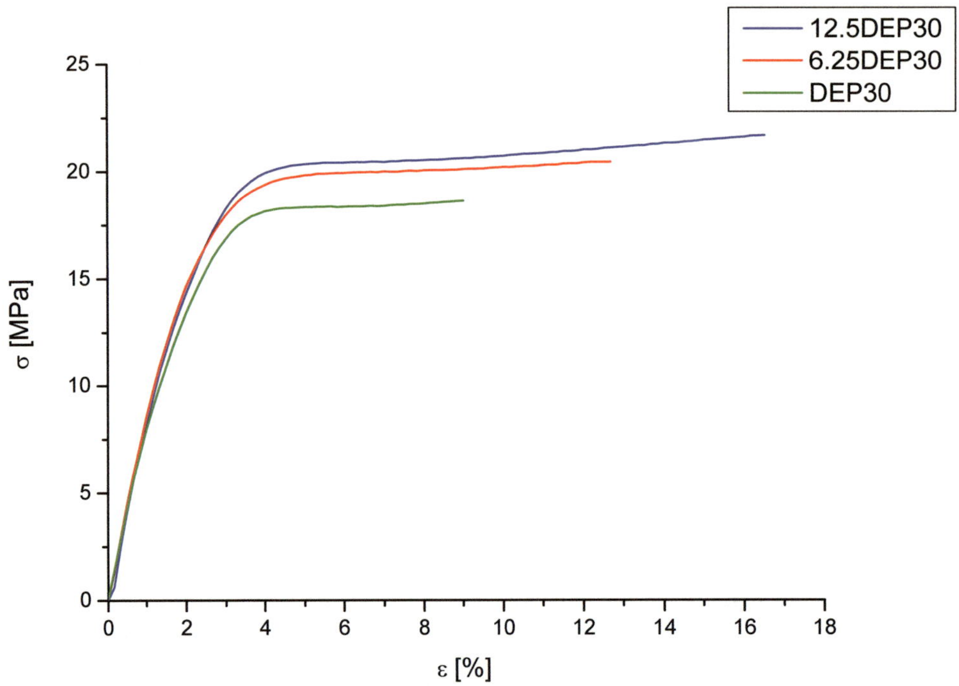

Fig. 11. The tensile stress-strain curves for samples prepared with 30% of DEP.

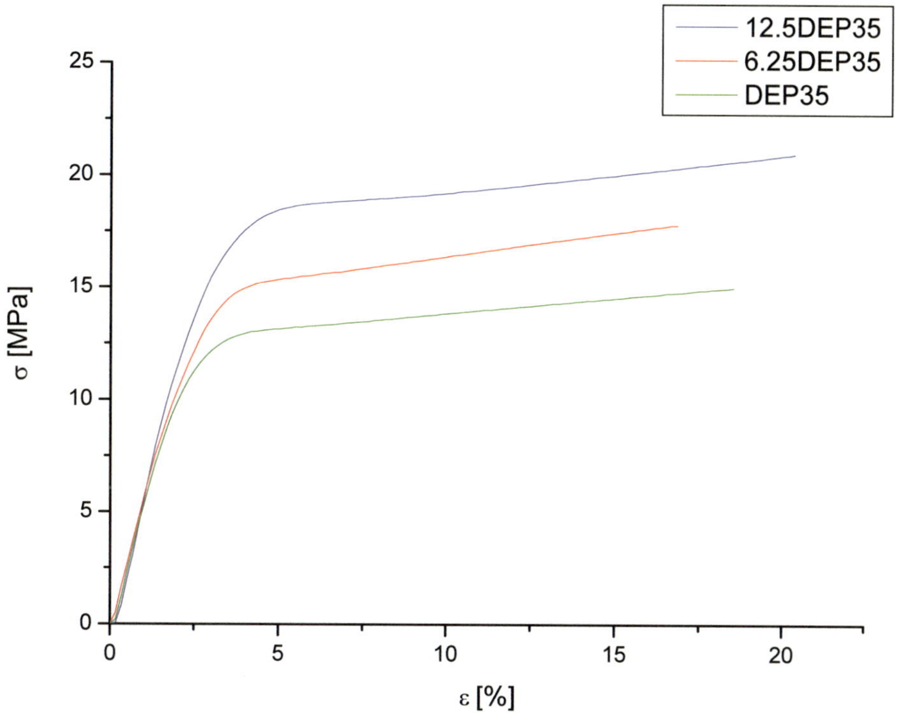

Fig. 12. The tensile stress-strain curves for samples prepared with 35% of DEP.

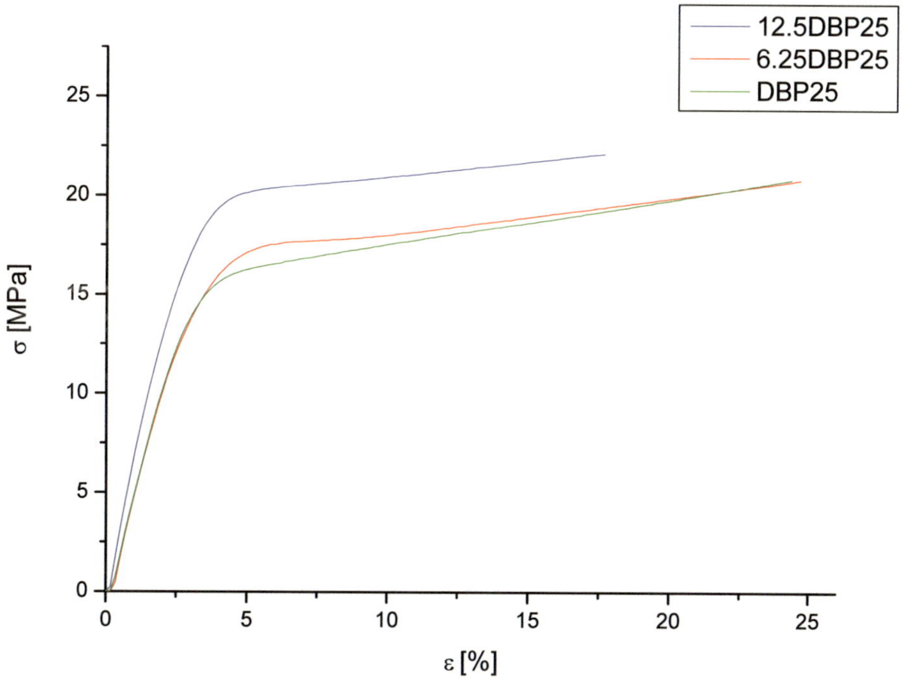

Fig. 13. The tensile stress-strain curves for samples prepared with 25% of DBP.

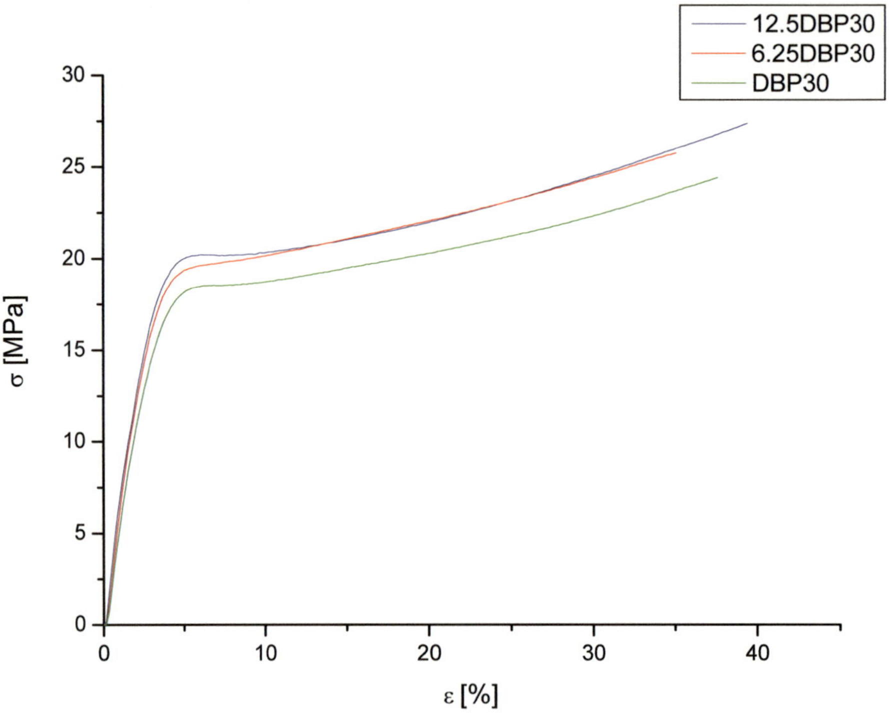

Fig. 14. The tensile stress-strain curves for samples prepared with 30% of DBP.

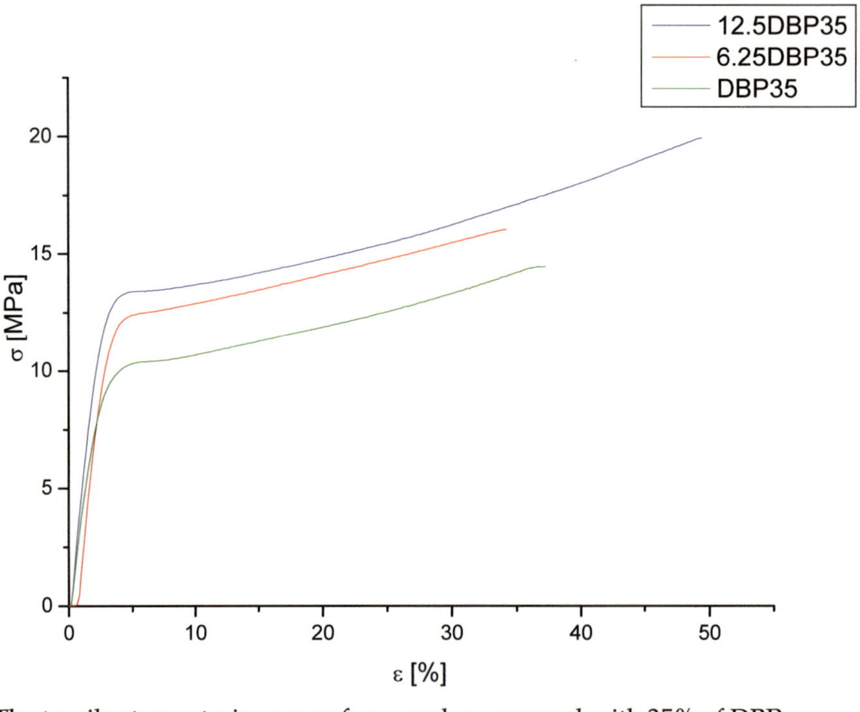

Fig. 15. The tensile stress-strain curves for samples prepared with 35% of DBP.

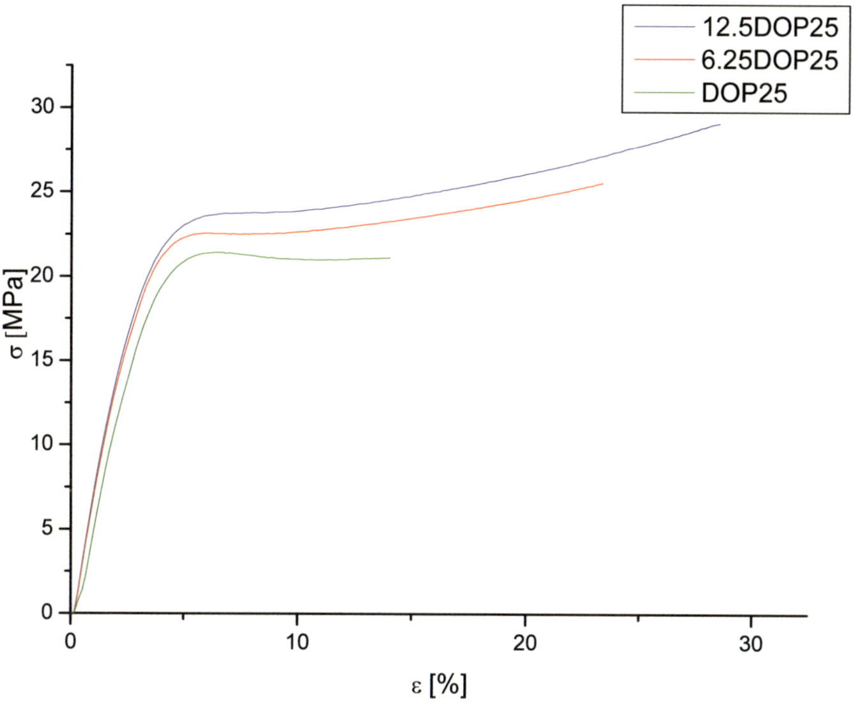

Fig. 16. The tensile stress-strain curves for samples prepared with 25% of DOP.

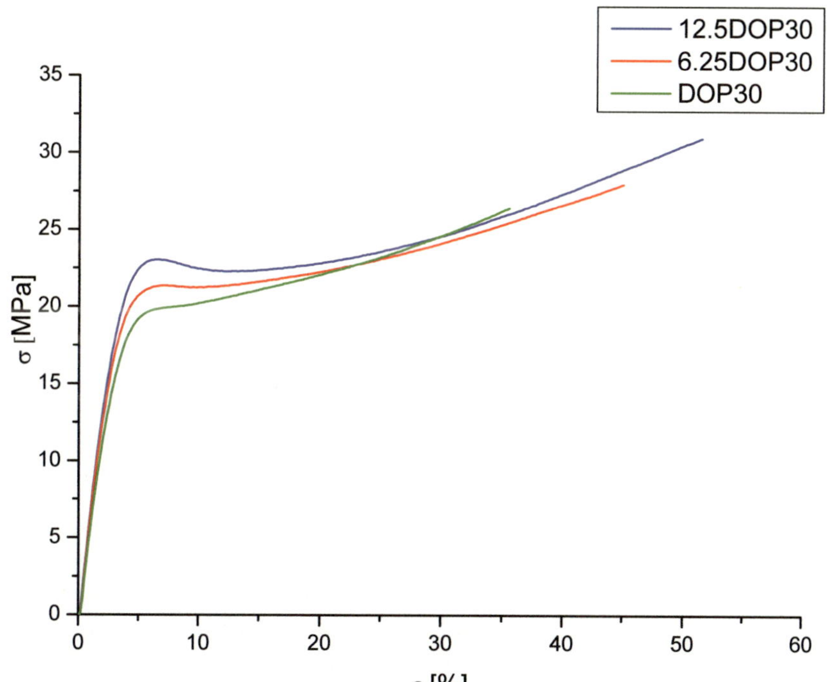

Fig. 17. The tensile stress-strain curves for samples prepared with 30% of DOP.

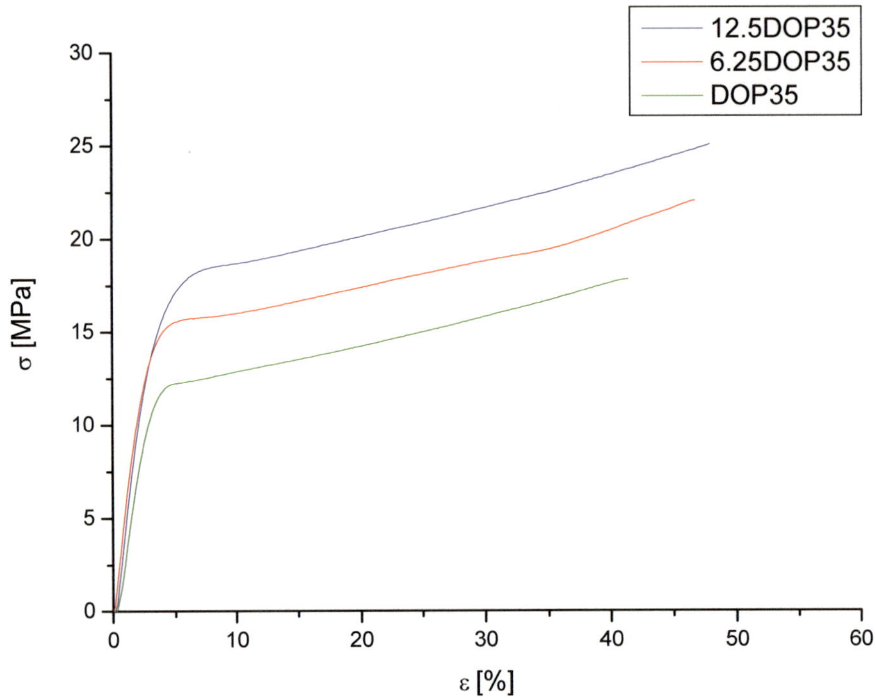

Fig. 18. The tensile stress-strain curves for samples prepared with 35% of DOP.

6. Conclusions

Taking into consideration obtained results we can conclude that type and amount of applied plasticizer as well as incorporation of inorganic phase into CAB matrix affected mechanical properties of the examined samples. Changing the type and concentration of the plasticizer, and amount of inorganic phase can modify the strength and extensibility of the materials. The higher the amount of incorporated silica, the harder and more brittle the material, however exhibiting good flexibility at 30 and 35% plasticizer concentration. All of the plasticizers investigated, excluding TEC, caused an antiplasticization effect at concentration 30% resulting in an increase in tensile strength, in comparison with the values at 25%. At higher concentration of plasticizers (35%) the additives caused plasticization reflected as a decreases in tensile strength and an increase in elongation at break values. Regarding the influence of inorganic phase incorporated into polymer matrix, the tensile strength was substantially improved, as compared with neat CAB, regardless of the plasticizer type.

Among all plasticizers, DEP was found to be the least efficient for CAB, as well as for organic-inorganic hybrids. Low plasticization efficiency showed also TEC. All samples prepared with DEP and TEC showed the noticeable low values of tensile strength as well as poor flexibility, as compared to the same formulations with other plasticizers used in this study. DOP, TBC and TEA were the most efficient plasticizers for CAB and organic-inorganic CAB hybrids. The best formulations in terms of mechanical properties were those containing 30% of above mentioned plasticizers. DOP at 30% concentration was the most effective to enhance the mechanical properties of CAB and organic-inorganic hybrids, with the highest tensile strength of 31.1 MPa for sample prepared from 87.5/12.5 polymer/TEOS formulation (12.5DOP30). Among citrate plasticizers used in this work, TBC, as well as TEA at 30% concentration were the most effective to improve mechanical properties.

As a final conclusion it can be stated that environmentally friendly citrate plasticizers can substitute phthalates in organic-inorganic CAB hybrids formulations. TBC and TEA can be used as valuable alternatives to DOP, producing materials displaying high values of tensile strength and satisfactory elongation at break.

7. References

[1] Ajayan P. M., Schadler L. S., Braun P. V., Nanocomposite Science and Technology, WILEY-VCH Verlag GmbH & Co. KGaA, Weinheim 2003.

[2] Kickelbick G. (Edit.), Hybrid Materials. Synthesis, Characterization, and Applications, WILEY-VCH Verlag GmbH & Co. KGaA, Weinheim 2007.

[3] Yano S., Iwata K., Kurita K., Physical properties and structure of organic-inorganic hybrid materials produced by sol-gel process, Materials Science and Engineering 1998, C6, p. 75-90.

[4] Kosaka P. M., Kawano Y., Petri H. M., Fantini M. C. A., Petri D. F. S., Structure and Properties of Composites of Polyethylene or Maleated Polyethylene and Cellulose or Cellulose Esters, Journal of Applied Polymer Science 2007, Vol. 103, p. 402-411.

[5] Benaniba M. T., Massardier-Nageotte V., Evaluation Effects of Biobased Plasticizer on the Thermal, Mechanical, Dynamical Mechanical Properties, and Permanence of Plasticized PVC, Journal of Applied Polymer Science 2010, Vol. 118, p. 3499-3508.

[6] Rahman M., Brazel Ch. S., The plasticizer market: an assessment of traditional plasticizers and research trends to meet new challenges, Progress in Polymer Science 2004, 29, p. 1223-1248.

[7] Vieira M. G. A., da Silva M. A., dos Santos L. O., Beppu M. M., Natural-based plasticizers and biopolymer films: A review, European Polymer Journal 2011, 47, p. 254-263.

[8] Han J. H. editor, Innovations in food packaging, Elsevier 2005, in Plasticizers in edible films and coatings Sothornvit R., Krochta J. M..

[9] Gil N., Saska M., Negulescu I., Evaluation of the effects of biobased plasticizers on the thermal and mechanical properties of poly(vinyl chloride), Journal of Applied Polymer Science 2006, vol. 102, p. 1366-1373.

[10] Wypych G. editor, Handbook of Plasticizers, ChemTec Publishing 2004.

[11] Elias H. G., An introduction to plastics, Second, completely revised edition, WILEY-VCH GmbH&Co. KGaA, Weinheim 2003.

[12] Ehrenstein G. W., Polymeric materials: structure, properties, applications, Carl Hanser Verlag, Munich 2001, chapter 4.2.2. Plasticization, p. 112-116.

[13] Zweifel H., Maier R. D., Schiller M., Plastics Additives Handbook, 6th edition, Carl Hauser Verlag, Munich 2009, chapter 3.13 Plasticizers.
[14] Daniels P. H., A Brief Overview of Theories of PVC Plasticization and Methods Used to Evaluate PVC-Plasticizer Interaction, Journal of Vinyl and Additive Technology 2009, Vol. 15, 4, p. 219-223.
[15] Wilkes Ch. E., Summers J. W., Daniels Ch. A. (Eds.), PVC Handbook, chapter 5 Plasticizers (L. G. Krauskopf, A. Godwin), Carl Hanser Verlag, Munich 2005.
[16] Plastic Additives Global Market to 2015 - Increasing Plastics Demand Supported by Recovering Global Economy Driving the Market, http://www.businesswire.com/news/home/20110221005492/en/Research-Markets-Plastic-Additives-Global-Market-2015.
[17] Craver C. D., Carraher C. E., Jr., Elsevier Science Ltd., (The Boulevard, Langford Lane, Kidlington Oxford 2000, UK, Polymer Science and Technology (section editor D. J. Lohse), chapter 9 (A. D. Godwin).
[18] Lindstrom A., Hakkarainen M., Environmentally Friendly Plasticizers for Poly(vinyl chloride)-Improved Mechanical Properties and Compatibility by Using Branched Poly(butylene adipate) as a Polymeric Plasticizer, Journal of Applied Polymer Science 2006, Vol. 100, p. 2180-2188.
[19] Eyerer P., Weller M., Hübner Ch. (Eds.), Polymers - Opportunities and Risks II: Sustainability, Product Design and Processing (The Handbook of Environmental Chemistry), Springer-Verlag, Berlin Heidelberg 2010, Additives for the Manufacture and Processing of Polymers, R. Höfer, K. Hinrichs.
[20] Market Study: Plasticizers, Ceresana Research, 2011, www.ceresana.com.
[21] Cao X. L., Phthalate Esters in Foods: Sources, Occurrence, and Analytical Methods, Comprehensive Reviews in Food Science and Food Safety 2010, Vol. 9, p. 21-43.
[22] Wittassek M., Koch H. M., Angerer J., Brüning T., Assessing exposure to phthalates – The human biomonitoring approach, Molecular Nutrition and Food Research 2011, 55, p. 7-31.
[23] Yin B., Hakkarainen M., Oligomeric Isosorbide Esters as Alternative Renewable Resource Plasticizers for PVC, Journal of Applied Polymer Science 2011, Vol. 119, p. 2400-2407.
[24] Babu B., Wu J. T., Biodegradation of phthalate esters by cyanobacteria, Journal of Phycology 2010, 46, p. 1106-1113.
[25] Imai Y., Kondo A., Iizuka H., Maruyama T., Kurohane K., Effects of phthalate esters on the sensitization phase of contact hypersensitivity induced by fluorescein isothiocyanate, Clinical and Experimental Allergy 2006, 36, p. 1462–1468.
[26] Persico P., Ambrogi V., Acierno D., Carfagna C., Processability and Mechanical Properties of Commercial PVC Plastisols Containing Low-Environmental-Impact Plasticizers, Journal of Vinyl and Additive Technology 2009, Vol. 15, 3, p. 139-146.
[27] Park H.-M., Misra M., Drzal L.T., Mohanty A.K., Green Nanocomposites from Cellulose Acetate Bioplastic and Clay: Effect of Eco-Friendly Triethyl Citrate Plasticizer. Biomacromolecules 2004, 5, p. 2281-2288.

[28] Mohanty A. K., Wibowo A., Misra M., Drzal L. T., Development of Renewable Resource-Based Cellulose Acetate Bioplastic: Effect of Process Engineering on the Performance of Cellulosic Plastics, Polymer Engineering and Science 2003, Vol. 43, No. 5, p. 1151-1161.

[29] Labrecque L. V., Kumar R. A., Dave V., Gross R. A., McCarthy S. P., Citrate Esters as Plasticizers for Poly (lactic acid), Journal of Applied Polymer Science 1997, Vol. 66, p. 1507-1513.

[30] Jiang L., Hinrichsen G., Biological degradation of cellulose acetate films: Effect of plasticizer, Die Angewandte Makromolekulare Chemie 1997, 253 p. 193-200.

[31] Wibowo A.C., Misra M., Park H.-M., Drzal L.T., Schalek R., Mohanty A.K., Biodegradable nanocomposites from cellulose acetate: Mechanical, morphological, and thermal properties, Composites Part A: Applied Science and Manufacturing 2006, 37, p. 1428-1433.

[32] Meier M. M., Kanis L. A., de Lima J. C., Pires A. T. N., Soldi V., Poly(caprolactone triol) as plasticizer agent for cellulose acetate films: influence of the preparation procedure and plasticizer content on the physico-chemical properties, Polymers for Advanced Technologies 2004, 15, p. 593-600.

[33] Ayuk J. E., Mathew A. P., Oksman K., The Effect of Plasticizer and Cellulose Nanowhisker Content on the Dispersion and Properties of Cellulose Acetate Butyrate Nanocomposites, Journal of Applied Polymer Science 2009,Vol. 114, p. 2723–2730.

[34] Wojciechowska P., Heimowska A., Foltynowicz Z., Rutkowska M., Degradability of organic-inorganic cellulose acetate butyrate hybrids in sea water, Polish Journal of Chemical Technology 2011, 13, 2, p. 29-34.

[35] Wojciechowska P., Foltynowicz Z., Polymer nanocomposites based on cellulose derivatives and their preparation, 2011, patent No. 209829, Polish Patent Office.

[36] Wojciechowska, P, Foltynowicz, Z. Synthesis of organic-inorganic hybrids based on cellulose acetate butyrate, Polimery 2009, 11–12, p. 845-848.

[37] Donempudi S., Yassen M., Controlled release PVC membranes: Influence of phthalate plasticizers on their tensile properties and performance, Polymer Engineering and Science 1999, Vol. 39, No. 3, p. 399-405

[38] Gutierrez-Villarreal M. H., Rodriguez-Velazquez J., The effect of citrate esters as plasticizers on the thermal and mechanical properties of poly(methyl methacrylate), Journal of Applied Polymer Science 2007, Vol, 105, p. 2370-2375.

[39] Zhang Y., Han J. H., Crystallization of High-Amylose Starch by the Addition of Plasticizers at Low and Intermediate Concentrations, Journal of Food Science 2010, Vol. 75, No. 1, p. 8-16.

[40] Vidotti S. E., Chinellato A. C., Hu G.-H., Pessan L. A., Effects of Low Molar Mass Additives on the Molecular Mobility and Transport Properties of Polysulfone, Journal of Applied Polymer Science 2006, Vol. 101, p. 825–832.

[41] Matuana L. M., Park Ch., B., Balatinecz J. J., The effect of low levels of plasticizer on the rheological and mechanical properties of Polyvinyl Chloride/Newsprint-Fiber Composites, Journal of Vinyl & Additive Technology 1997, Vol. 3., No. 4, p. 265-273.

[42] Moraru C. I., Lee T.-C., Karwe M. V., Kokini J. L., Plasticizing and Antiplasticizing Effects of Water and Polyols on a Meat-Starch Extruded Matrix, Journal of Food Science 2002, Vol. 67, Nr. 9, p. 3396-3401.

Characterization of High Molecular Weight Poly(vinyl chloride) – Lithium Tetraborate Electrolyte Plasticized by Propylene Carbonate

Ramesh T. Subramaniam[1,*], Liew Chiam-Wen[1], Lau Pui Yee[2] and Ezra Morris[2]
[1]Centre for Ionics University Malaya, Department of Physics, Faculty of Science, University of Malaya, Kuala Lumpur
[2]Faculty of Engineering and Science, Universiti Tunku Abdul Rahman, Kuala Lumpur
Malaysia

1. Introduction

An electrolyte is a substance consisting of free ions and acts as a medium channel for transferring the charges between a pair of electrodes. Sometimes, they are also referred as "lytes" which is derived from the Greek word, "lytos", which means "it may be dissolved". Electrolyte is comprised of positively charged species which is called as cation and negatively charged species, anion. The properties of an electrolyte can be exploited via electrolysis process: separation of chemically bonded element or compounds by applying the electrical current. In the early stage, liquid electrolytes have been discovered and investigated. Liquid electrolyte is a substance that conducts the electricity in an aqueous solution by migrating both cations and anions to the opposite electrodes through an electrically conducting path as a useful electric current. However, it faces problem of leakage of hazardous liquids or gases. Other drawbacks are formation of lithium dendrite, electrolytic degradation of electrolyte and uses of flammable organic solvent (Ramesh et al., 2011a). Apart from that, it exhibits poor long-term stability due to the evaporation of the liquid phase in the cells (Yang et al., 2008). Therefore, solid polymer electrolytes (SPEs) were synthesized to prevail over the limitations of conventional liquid electrolytes.

1.1 Solid polymer electrolytes

The development on SPEs was initiated by the pioneering work of Wright et al. three decades ago (Lee and Wright, 1982). A polymer electrolyte (PE) is defined as a solvent-free system whereby the ionically conducting pathway is generated by dissolving the low lattice energy metal salts in a high molecular weight polar polymer matrix with aprotic solvent (Gray, 1997a). The fundamental of ionic conduction in the polymer electrolytes is the covalent bonding between the polymer backbones and ions. Initially, the electron donor group in the polymer would form solvation onto the cation component in the doping salt and thus facilitate the ion separation, leading to ionic hopping mechanism. Hence, it generates the ionic conductivity. In other words, the ionic conduction of PE arises from rapid segmental motion of polymer matrix combined with strong Lewis-type acid-base interaction between the cation and donor atom (Ganesan et al., 2008). However, the well separated ions might be poor conductors if the ions are immobile and unable for the migration. Therefore, the host polymer must be sufficiently flexible to provide enough space for the migration of these two ions.

SPE serves three principal roles in a lithium rechargeable battery. Firstly, it acts as the electrode separator that insulates the anode from the cathode in the battery which removes the requirement of inclusion of inert porous membrane between the electrolytes and electrodes interface. Besides, it plays the role as medium channel to generate ionic conductivity which ions are transported between the anode and cathode during charging and discharging. This induces to enhancement of energy density in the batteries with formation of thin film. In addition, it works as binders to ensure good electrical contact with electrodes. Thus, high temperature process for conventional liquid electrolytes is eliminated as well (Gray, 1991; Kang, 2004).

1.2 Applications of solid polymer electrolytes

SPEs are of great interests in the technology field, especially the area of electrical power generation and storage systems. It is primarily due to their wider range of applications, ranging from small scale production of commercial secondary lithium ion batteries (also known as the rechargeable batteries) to advanced high energy electrochemical devices, such as chemical sensors, fuel cells, electrochromic windows (ECWs), solid state reference electrode systems, supercapacitors, thermoelectric generators, analog memory devices and solar cells (Gray, 1991; Rajendran et al., 2004). As for the commercial promises of lithium rechargeable batteries, there is a wide range of applications which ranges from portable electronic and personal communication devices such as laptop, mobile phone, MP3 player, PDA to hybrid electrical vehicle (EV) and start-light-ignition (SLI) which serves as traction power source for electricity (Gray, 1997a; Ahmad et al., 2005).

1.3 Advantages of solid polymer electrolytes

A force had been driven in the development of SPEs to replace conventional liquid electrolytes due to its attractive advantages. These features include the elimination of the problems of corrosive solvent leakage and harmful gas production, ease of processability due to elimination of liquid component, and suppression of lithium dendrite growth as well (Rajendran et al., 2004; Ramesh et al., 2010). Besides safety performance, SPE is also a promising candidate because of its intrinsic characteristics, such as ease to configure in any shape due to its high flexibility of polymer matrix, high automation potential for electrode application and no new technology requirement as well as light in weight (Xu and Ye, 2005; Gray, 1991). The other advantages of SPEs are viz., negligible vapor pressure, ease of handling and manufacturing, wide operating temperature range, low volatility, high energy density and high ionic conductivity at ambient temperature (Baskaran et al., 2007; Rajendran et al., 2004). In addition, they exhibit excellent electrochemical, structural, thermal, photochemical and chemical stabilities without the combustible reaction products at the electrode surface by comparing with conventional liquid electrolyte (Adebahr et al., 2003; Nicotera et al., 2006; Stephan, 2006). Moreover, they produce miniaturized structures via fabrication methods and provide longer shelf-lives with no internal shorting (Gray, 1997a; Stephan, 2006).

1.4 Gel polymer electrolytes (GPEs)

SPEs possess high mechanical integrity; however they exhibit low ionic conductivity. Therefore, gel polymer electrolytes (GPEs), sometimes known as gelionic solid polymer electrolytes are yet to be developed to substitute SPEs because of its inherent characteristics (Stephan et al., 2000a). Such features are low interfacial resistance, decrease in reactivity, improved safety and exhibit better shape flexibility as well as significant increases in ionic conductivity with a small portion of plasticizers (Ahmad et al., 2008; Pandey and Hashmi, 2009). GPE is obtained by dissolving the host polymer along with a metal dopant salt in a polar organic solvent (more commonly known as plasticizer) (Osinska et al., 2009; Rajendran et al., 2008a). In other words, it is an immobilization of a liquid electrolyte in a polymer matrix (Han et al., 2002). As a unique characteristic, GPEs possess both cohesive properties of solids and the diffusive property liquids. Even though they are in a solid state, but at atomic level, the local relaxations provide liquid-like degree of freedom which is comparable to those conventional liquid electrolytes. Moreover, GPEs show better mechanical and electrochemical properties within a wide operational temperature range in comparison with that of liquid electrolytes (Ahmad et al., 2005; Stephan et al., 2000a). Other attractive advantages are leak proof construction, lighter, cheaper and easy fabrication into desired shape and size (Zhang et al., 2011). They also maintain the interfacial contacts under stresses such as the changing of volume associated with cell charging and discharging. GPEs could form good interfacial contacts with electrode materials as they are not brittle as solid crystalline or glass electrolytes (Gray, 1997a).

2. Methods to improve ionic conductivity

Ionic conductivity is the main aspect to be concerned in the solid polymer electrolytes. Ionic conductivity is defined as ionic transportation from one site to another through defects in the crystal lattice of a solid under the influence of an external electric field. Much effort has been devoted for developing highly conducting polymer electrolytes. Several ways have been done to modulate the ionic conductivity of polymer electrolytes, for instance, random and comb-like copolymer of two polymers, polymer blending, mixed salt and mixed solvent systems as well as impregnation of additives such as ceramic inorganic fillers and plasticizers.

2.1 Random copolymers

In the first study of poly(ethylene oxide) (PEO) with various inorganic lithium salts, PEO showed low conductivity at ambient temperature due to the higher degree of crystallization. To overcome this obstacle, reduced crystallinity or amorphous polyether-based host architectures have been focused. Random copolymerization is one of routes to produce amorphous host polymers. Booth and co-workers had successfully synthesized random oxyethylene-oxymethylene polymer structures in year 1990. Ethylene oxide monomers are randomly interspersed with methylene oxide groups. Thus, the methylene oxide would break up the regular helical structure of PEO and suppress the crystalline region of polymer matrix. A similar random copolymer had been synthesized by replacing the methylene oxide with dimethyl siloxy units. This resultant polymer matrix illustrated higher flexibility which assists in ionic conduction (Gray, 1997b).

2.2 Comb polymers

In general, comb polymers contain pendant polymer chain and they are structurally related to grafting copolymers. The comb-branched system containing of low molecular weight of polyether chain is grafted to a polymer backbone. Thus, it lowers the glass transition temperature (T_g) and then helps to optimize the ionic conductivity by improving the flexibility of polymer chain into the system. The elastic poly[ethylene oxide-co-2-(2-methoxyethoxy)ethyl glycidyl ether] [P(EO/MEEGE)]-based polyether comb polymer electrolytes were synthesized by Nishimoto and co-workers. The degree of crystallinity was decreased with increasing the composition of MEEGE in copolymers, which in accordance with higher ionic conductivity. The introduction of the side chain of MEEGE in the copolymers enhances the flexibility of polymer matrix and hence improves the ion mobility. The highest ionic conductivity of 10^{-4} Scm^{-1} was achieved at room temperature (Nishimoto et al., 1998). Until today, this technique is still being employed in this area. Recently, many researchers have keen of interest on polyvinylidene-co-hexafluorophosphate (PVdF-co-HFP) copolymer. Composite polymer electrolytes prepared by adding SiO_2 nanowires into (PVdF-co-HFP) are described by Zhang et al. The ionic conductivity of this composite polymer electrolyte is up to 1.08×10^{-3} Scm^{-1} with the electrochemical window of 4.8 V (Zhang et al., 2011).

2.3 Polymer blending

Polymer blending is physical means to mix two or more different polymers or copolymers which are not linked by covalent bonds. This polymer blend is a new macromolecular material with special combinations of properties. For polymer blends, a first phase is adopted to absorb the electrolyte active species, whereas the second phase is tougher and sometimes substantially inert. It is a feasible way to increase the ionic conductivity because it offers the combined advantages of ease of preparation and easy control of physical properties within the definite compositional change (Rajendran et al., 2002). Polymer blending is of great interest due to their advantages in properties and processability compared to single component. In industry area, it enhances the processability of high temperature or heat-sensitive thermoplastic in order to improve the impact resistance. Besides, it can reduce the cost of an expensive engineering thermoplastic. The properties of polymer blends depend on the physical and chemical properties of the participating polymers and on the state of the phase, whether it is in homogenous or heterogeneous phase. If two different polymers able to be dissolved successfully in a common solvent, this polymer blends or intermixing of the dissolved polymers will occur due to the fast establishment of the thermodynamic equilibrium (Braun, 2005). Sivakumar et al. (2006) observed that PVA (60 wt%)-PMMA (40 wt%)-$LiBF_4$ complex exhibits the maximum conductivity of 2.8×10^{-5} Scm^{-1} at ambient temperature. It is also higher than the pure PVA system which has been reported to be 10^{-10} Scm^{-1}.

2.4 Mixed salt system

The conductivity of the mixed salts in polymer electrolyte is higher than single salt electrolyte. It is due to the addition of second salt may prevent the formation of aggregates and clusters. Therefore, it increases the mobility of ion carriers (Gray, 1997b). An approach had been done by Ramesh and Arof (2000). In this research, we synthesized poly (vinyl chloride) (PVC)-based polymer electrolytes with lithium trifluoromethanesulfonate (LiTf) and lithium tetrafluoroborate ($LiBF_4$) as doping salts. The ionic conductivity is increased by four orders of magnitude in comparison with single salt system. It is attributed to the increase in the mobility of charge carriers by avoiding the aggregation process.

2.5 Mixed solvent system

On the other hand, the increase of conductivity in binary solvent system is proven by Deepa et al. (2002). In this study, poly(methyl methacrylate) (PMMA)-based polymer electrolytes containing lithium perchlorate ($LiClO_4$), with a mixture of solvents of propylene carbonate (PC) and ethylene carbonate (EC) were prepared. The maximum ionic conductivity of 10^{-3} S cm^{-1} was obtained and it was increased by two orders of magnitude as compared to polymer electrolyte system with single solvent. Synergistic effect is the major factor to increase the ionic conductivity in this mixed solvent system. Different physicochemical properties of the individual solvents come into play and contribute to high ionic conductivity in the presence of the effect. The preparation on EC/PC/2-methyl-tetrahydrofuran (2MeTHF) ternary mixed solvent electrolyte had been done by Tobishima and co-workers. They found out that the discharge capacity of Li/amorphous V_2O_5-P_2O_5 cells with a ternary mixed solvent electrolyte are slightly better than for cells with EC/PC binary mixed solvent electrolytes.

2.6 Addition of inorganic fillers

Utilization of common additives such as inorganic fillers and plasticizers is the effective and efficient approach to enhance the ionic conductivity. Fillers (also known as reinforcing fillers) are divided into two types: inorganic and organic. Variety types of inorganic fillers have been used, including mica, clay, titania (TiO_2), fumed silica (SiO_2) and alumina (Al_2O_3). On the other hand, graphite fibre and aromatic polyamide are some examples of organic fillers. The main objectives of dispersion of inorganic filler are to alter the properties of the polymer, enhance processability and improve the mechanical stability in the polymer electrolyte system. Dispersion of inorganic fillers can also improve the ionic conductivity in the polymer electrolyte. Besides improving the lithium transport properties, the inclusion of ceramic filler has been found to enhance the interfacial stability of polymer electrolytes (Osinska et al., 2009).

2.7 Plasticization

A number of attempts have been made on plasticized-polymer electrolytes in order to rise up the ionic conductivity greatly. Plasticization is generally recognized as one of the effective and efficient methods available for decreasing the crystalline region of polymer electrolytes (Suthanthiraraj et al., 2009). Plasticizer is a non-volatile and low molecular weight aprotic organic solvent which has a T_g in the vicinity of -50 °C. Carbonate ester such as propylene carbonate (PC), ethylene carbonate (EC), dimethyl carbonate (DMC) and diethyl carbonate (DEC), and high dielectric constant solvent such as N,N-dimethylformamide (DMF), N,N-dimethylacetamide (DMAc) and γ-butyrolactone are widely used as main components of GPE (Pradhan et al., 2005; Suthanthiraraj et al., 2009; Ning et al., 2009). Other examples of common plasticizers are dibuthyl phthalate (DBP), diocthyl adipate (DOA) and polyethylene glycol (PEG) (Suthanthiraraj et al., 2009).

2.7.1 Advantages of plasticizers

The effect of plasticizers on the polymer electrolytes entirely depends on the specific characteristics of the plasticizer, for example, viscosity, dielectric constant, the interaction between polymer and plasticizer, and the coordinative bond between ion and plasticizer (Rajendran and Sivakumar, 2008b). The incorporation of plasticizer not only enhances the salt solvating power, but it also provides sufficient mobility of ions with a better contact between polymer electrolytes and electrodes (Ramesh and Arof, 2001; Rajendran et al., 2004). Apart from that, plasticizer is an attractive additive due to its superior miscibility with polymer, high dielectric constant and low viscosity (Ramesh and Chao, 2011b). Inclusion of plasticizer is the most successful skill to enhance the ionic conductivity without compromising the thermal, electrochemical and dimensional stabilities (Ganesan et al., 2008). Plasticizer are expected to improve the ionic conductivity through some important intrinsic modifications such as significant changes in local structure, enhancement of amorphous fraction and changes in local electric field distribution in the polymer matrix.

2.7.2 Roles of plasticizers

The principal function of a plasticizer is to reduce the modulus of polymer at the desired temperature by lowering its T_g. In this theory, the increase in concentration of plasticizer causes the transition from the glassy state to rubbery region at progressively lower temperature. Moreover, it reduces the viscosity of polymer system and then facilitates the ionic migration within the polymer matrix. Besides, it weakens the interactions within the polymer chains and thus improves the flexibility of polymer chains in the polymer matrix (Ganesan et al., 2008). As a result, it increases the free volume of polymer and enhances the long-range segmental motion of the polymer molecules in the system. In an approach, the polymer matrix is swollen in a plasticiser, the latter being an aprotic solvent with a high dielectric constant. A new mobile pathway for ion migration is being introduced upon addition of plasticizer by dissociating the charge carriers. Hence, it increases the amount of mobile charge carriers and promotes the ionic transportation, enhancing the ionic conductivity. The polymer component would, on the other hand, render necessary stability to the lithium anode electrolyte interface, which thereby reduces the chance of dendrite growth on the lithium anode (Rajendran et al., 2000a). In general, plasticizers are having conjugated double bond which initializes the delocalization of electrons. Therefore, it improves the donor capacity of oxygen atom which facilitates the binding of cations.

2.7.3 Literature review of plasticized-gel polymer electrolytes

As reported in Michael et al. (1997), three types of ester class plasticizers, namely dioctyl phthalate (DOP), dibutyl phthalate (DBP) and dimethyl phthalate (DMP), were employed to examine its effect on ionic conductivity in the PEO-$LiClO_4$ polymer complex. Among these plasticizers, DOP was found to be the excellent plasticizer in term of thermal stability as proven in differential thermal analysis (DTA). The results show the reduced weight loss as increases the plasticizer concentration. Ali et al. (2007) studied the plasticized-polymer electrolytes that composed of PMMA, propylene carbonate (PC) or ethylene carbonate (EC) as plasticizer and LiTf or $LiN(CF_3SO_2)_2$ as dopant salt. According to this literature, the ionic conductivity increases with the concentration of the plasticizer. They also declare that the PC-based plasticized-polymer electrolytes exhibit higher ionic conductivity compared to the EC-based plasticized-polymer electrolytes (Ali et al., 2007).

Rajendran et al. (2004) also incorporated few types of plasticizers in the polymer electrolytes containing PVA/PMMA-$LiBF_4$. The highest ionic conductivity of 1.29 $mScm^{-1}$ had been observed for EC-based complex because of higher dielectric constant of EC (ε=85.1) (Rajendran et al., 2004). The addition of the plasticizer has been reported to reduce the crystallinity of the polymer complexes which in turns to a better ionic conductivity (Kelly et al., 1985). Kelly and co-workers assert that the presence of plasticizer exhibits downward shift in T_g due to the weaker interactions between the ions and polymer chain which in accordance with higher ion dissociation. A maximum electrical conductivity of 2.60×10^{-4} Scm^{-1} at 300 K has been observed for 30wt.% of PEG as plasticizer compared to the pure PEO-$NaClO_4$ system of 1.05×10^{-6} Scm^{-1}. This can be explained that the addition of plasticizer enhances the amorphous phase in with concomitant the reduction in the energy barrier. Eventually, it results higher segmental motion of lithium ions (Kuila *et al.*, 2007).

3. Materials

In this study, poly (vinyl chloride) (PVC), lithium tetraborate ($Li_2B_4O_7$) and propylene carbonate (PC) were employed as host polymer, dopant salt and plasticizer, respectively. In this section, we discuss about the general description of the materials and the reasons for choosing the materials.

3.1 Poly(vinyl chloride) PVC

Apart from PEO, poly(vinyl alcohol) (PVA), poly(acrylonitrile) (PAN), poly(ethyl methacrylate) (PEMA), poly(vinyl chloride) (PVC), poly(vinylidene fluoride) (PVdF) have also been used as polymer host materials. PVC is a thermoplastic polymer where its IUPAC name is poly(chloroethanediyl). It consists of numerous repeating units of monomers called vinyl chloride. It is a vinyl polymer composing of numerous repeating units of CH_2-CHCl.

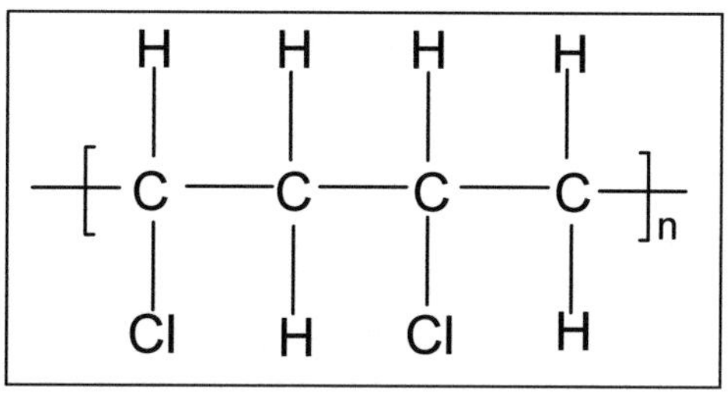

Fig. 1. Chemical structure of PVC.

PVC is mainly produced by radical polymerization (Endo K, 2002). In this polymerization, it associates the vinyl chloride molecules and thus forms the polymeric chains of macromolecules. From the scientific point of view, lone pair of electrons at the chlorine atoms is the main reason for choosing PVC as host polymer. Thus, it can form solvation onto lithium salts easily (Ramesh and Chai, 2007). PVC is chosen due to its high compatibility with the liquid electrolyte, good ability to form homogeneous hybrid film, commercially available and inexpensive (Li et al., 2006). Other unique characteristics are easy processability and well compatible with a large number of plasticizers (Ramesh and Ng, 2009). It plays an important role as mechanical stiffener because of the dipole-dipole interactions between the hydrogen and chlorine atoms (Ramesh and Chai, 2007).

3.2 Lithium tetraborate ($Li_2B_4O_7$)

Several types of lithium salt has widely been used, such as lithium hexaflurorophosphate ($LiPF_6$), lithium hexafluoroarsenate ($LiAsF_6$), lithium bis(trifluoromethanesulfonyl) imide (LiTFSI), LiTf, $LiClO_4$ and $LiBF_4$. Lithium tetraborate (also known as boron lithium oxide or dilithium borate) (LBO or more commonly known as $Li_2B_4O_7$) was employed in this study. This compound is generally defined as one type of dopant used to provide lithium cations in the preparation of polymer electrolytes. It is constructed of lithium cations and tetraborate anions, where its stoichiometric ratio is two cations to one anion, as illustrated as below.

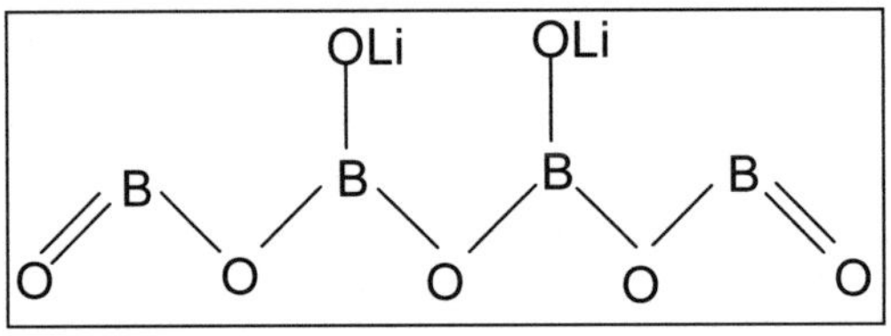

Fig. 2. Chemical structure of lithium tetraborate.

In addition, it is a new non-ferroelectric piezoelectric substrate material with a congruent melting point of 917 °C. LBO single crystal is also a superior substrate for surface acoustic wave (SAW) and bulk acoustic wave (BAW) devices as it has a specific crystallographic plane (110). Thus, it provides a zero temperature coefficient of frequency and a fairly high electromechanical coupling factor. According to Byrappa and Shekar (1992), LBO occurs naturally as diomignite, i.e. colourless crystals in fluid inclusions in the mineral spodumene. It appears as a loosely attached crystal with a very fine-grained structure. These intrinsic and unique properties make it as excellent material by comparing with other lithium salt. The fine structure of $Li_2B_4O_7$ would enhance the solubility in the polymer matrix and eventually, speed up the salt dissociation process. Other factors to choose it as doping salt are abundant availability of raw materials and no environmental pollution (Xu et al., 2004). As aforementioned, $Li_2B_4O_7$ is naturally obtained from mineral. Therefore, it is a cost effective material compared to those synthetic lithium salts, for instance $LiBF_4$, LiTFSI and LiTf. High toxicity of $LiAsF_6$ and poor chemical and thermal stabilities of $LiPF_6$ are not good choices as doping salt in polymer electrolyte. Likewise, $LiClO_4$ reacts with most organic species readily in violent ways under certain conditions such as high temperature and high current charge because of high oxidation state of chlorine (VII) in perchlorate. Moreover, the corrosion of a key component of the cell by TFSI anions restricts the possible application of LiTFSI greatly in the polymer electrolytes. Poor ionic conductivity of LiTf in non-aqueous solvent which caused by its low dissociation constant in low dielectric media and its moderate ion mobility, is the major shortcoming of LiTf as compared with other lithium salts (Kang, 2004). Therefore, it can be concluded that $Li_2B_4O_7$ is an indispensable electrolyte solute as it shows multiple merits than other lithium salts.

3.3 Propylene carbonate (PC)

As aforesaid, plasticizer could enhance the ionic conductivity of polymer electrolytes. Propylene carbonate (PC) is used in this study. PC is an organic, colourless and odourless organic compound. It is also well known as highly polar and aprotic solvent. Furthermore, it is a byproduct of the synthesis of polypropylene carbonate from propylene oxide and carbon dioxide. It can be obtained from the synthesis of urea and polypropylene glycol in the presence of zinc-iron double oxide catalysis. It is composed of twofold ester of propylene glycol and carbonic acid as illustrated as below.

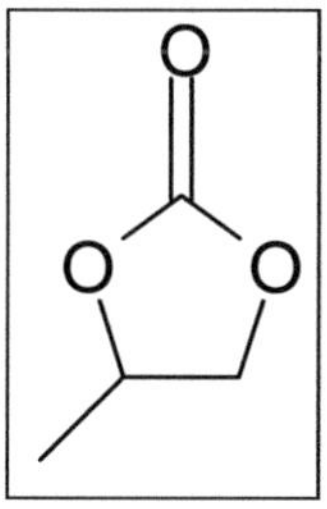

Fig. 3. Chemical structure of propylene carbonate.

PC is a preferred solvent as it exhibits many unique characteristics. These properties are static stability with lithium, wide liquid range and low freezing point (Kang, 2004). Many researchers have drawn interests onto this plasticizer due to its high dielectric constant (ε= 64.9). This high dielectric constant could help in the dissociation of the charge carriers, especially the cations from the dopant salt. The electrical performance would be improved as it manifests excellent plasticizing effect (Tobishima and Yamaji, 1984). Cations are more readily to be obtained as the salt is dissolved in PC through electrolysis process, because of its high molecular dipole moment of 4.81 D (Jorné and Tobias, 1975). Moreover, high polarity of this plasticizer creates an effective solvation shell around the cations and hence forms a conductive electrolyte. PC displays the highest dielectric constant in comparison with *DEC* (ε= 2.8), *ethyl methyl carbonate (EMC)* (ε= 2.9), *DMC* (ε= 3.1), DOA (ε= 4–5), DBP (ε= 6.4), DMP (ε= 8.5) and benzyl acetate (BC) (ε= 53) (Kang, 2004; Gu et al., 2006). Despite EC illustrates higher dielectric constant than PC, however PC exhibits wider range of liquidity than EC as its melting point is up to −48.8 °C. In contrast, high melting temperature of 36.4 °C is the main shortcoming of EC. High dipole moment of PC than EC (4.61 D) with low vapor pressure also compensates the obstacle of lower dielectric constant of PC (Kang, 2004). Therefore, PC is becoming an attractive prospect as plasticizer compared to other plasticizers.

3.4 Tetrahydrofuran (THF)

Tetrahyrdofuran (THF) is a colorless, water-miscible organic liquid with low viscosity at standard temperature and pressure. It is one of the most polar ethers with a wide liquid range and is widely been used as solvent. In addition, it is an aprotic and highly volatile solvent with dielectric constant of 7.6. Another virtue of this solvent is its good solvent properties. It can dissolve a wide range of non-polar and polar chemical substances. Therefore, it is being chosen as solvent as it shows a well-dissolution with PVC, $Li_2B_4O_7$ and PC in this study. It is also a heterocyclic compound with chemical formula of $(CH_2)_4O$, as shown in Figure 4.

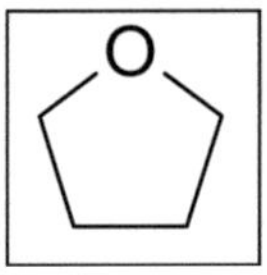

Fig. 4. Chemical structure of tetrahydrofuran.

4. Experimental

These plasticized-polymer electrolytes were prepared by solution casting technique. It is the simplest and cost-effective method to produce thin films from solution by evaporating the volatile solvent. No technology is required in this technique.

4.1 Materials

High molecular weight poly (vinyl chloride) PVC was obtained from Fluka, while lithium tetraborate ($Li_2B_4O_7$) and propylene carbonate (PC) were obtained from Aldrich. Tetrahydrofuran (THF) was obtained from J.T. Baker.

4.2 Sample preparation

Prior to the preparation of the polymer electrolytes, $Li_2B_4O_7$ was dried at 100 °C for 1 hour to eliminate trace amounts of water in the material. Appropriate amount of PVC, $Li_2B_4O_7$ and PC were dissolved in anhydrous THF. The mixture was then stirred continuously for 24 hours at room temperature to achieve a homogenous solution. The resulting solution was cast on a Petri dish and was allowed to evaporate slowly in a fume hood. This procedure yields free standing films eventually. The designations of polymer film are listed in Table 1. The weight ratio of PVC (70%) to $Li_2B_4O_7$ (30%) had achieved the maximum ionic conductivity in the preliminary step which is not being shown here. As a result, the ratio of PVC to $Li_2B_4O_7$ was fixed as 70 wt% to 30 wt% in this study.

Designation	Composition of materials [PVC: $Li_2B_4O_7$: PC (wt%)]
SPC1	70.0:30.0:0.0
SPC2	66.5:28.5:5.0
SPC3	63.0:27.0:10.0
SPC4	59.5:25.5:15.0
SPC5	56.0:24.0:20.0
SPC6	52.5:22.5:25.0
SPC7	49.0:21.0:30.0
SPC8	45.5:19.5:35.0
SPC9	42.0:18.0:40.0

Table 1. The nomenclature of samples with different stoichiometric amounts of materials added into polymer electrolytes.

4.3 Instrumentation

After the samples were prepared, the characterizations have been employed to investigate the electrical, structural and thermal properties of the samples. These analytical and evaluative methods include ac-impedance spectroscopy, Fourier Transform Infrared spectroscopy (FTIR) and thermogravimetry analysis (TGA).

4.3.1 Ac-impedance spectroscopy

Impedance spectroscopy (IS) is a powerful analytical tool to characterize the electrical properties of materials and their interfaces with electronically conducting electrodes. It is also widely been used to envisage the dynamics of bound or mobile charge in the bulk or interfacial regions of any kind of solid or liquid material: ionic, semiconducting, mixed electronic-ionic and even insulator (dielectric) (Barsoukov & Macdonald, 2005). The principle of the impedance spectroscopy is based on the ability of a medium to pass an alternating electrical or frequency current. It is well functioned by conducting current and measuring the potential difference created by the circulation of this current. When an electric field is applied across the sample, the polar group might be activated as dipoles which always interact with the corresponding ions due to the Coulombic electric force. Thus, these dipole moments will rearrange themselves under the influence of the external electric field, depending on the mobility of backbone. So, lithium cations can travel faster along these activated or polarizing areas to reach opposite of the electrode and generate current (Selvasekarapandian et al., 2006).

The prepared samples were subjected to ac-impedance spectroscopy. The thickness of the samples was measured by using micrometer screw gauge. The ionic conductivities of the samples were determined, by using HIOKI 3532-50 LCR HiTESTER, over a frequency range between 50 Hz and 1 MHz. The ionic conductivity was measured from ambient temperature to 100 °C. Samples were mounted on the holder with stainless steel (SS) blocking electrodes under spring pressure with the configuration SS/SPE/SS.

4.3.2 Fourier Transform Infrared Spectroscopy (FTIR)

The main fundamental of FTIR is to determine structural information about a molecule. The main principle of FTIR is related to the interferometry which is an optic study. It separates infrared beam of light which serves as light source radiation into two ray beams. Once the beam of infrared is passed through the sample, the molecules would absorb the infrared radiation and then excite to a higher energy state. Thus, the energies associated with these vibrations are quantized; within a molecule, only specific vibrational energy levels are allowed. The amount of energy absorbed at each wavelength was recorded. The frequencies which have been absorbed by the sample are determined by detector and the signal is amplified. Hence, IR spectrum was obtained. FTIR spectroscopy is not only applied in the crystalline region of complexation, whereas the complexation in amorphous phase can also be determined.

FTIR analysis was performed by using Perkin-Elmer FTIR spectroscopy RX 1 in the wave region between 4000 and 400 cm^{-1}. The resolution of the spectra was 4 cm^{-1} and recorded in the transmittance mode.

4.3.3 Thermogravimetry Analysis (TGA)

TGA is a versatile thermal study in polymer field. It is primarily used to determine thermal stability and thermal degradation of the samples as a function of change in temperature under inert conditions. The main principle of TGA is to monitor the weight of the samples on a sensitive balance (also known as thermobalance) continuously as the sample temperature is increased, under an inert atmosphere or air at a controlled uniform rate. The data were recorded as a thermogram of weight which is in y–axis against sample temperature which is in x–axis.

The thermal stability of polymer films was performed by Mettler Toledo Thermal Gravimetric Analyser which comprised of TGA/SDTA851@ as main unit and STARe software. Sample weighing 2–3 mg placed into 150 µl of silica crucible. The samples were then heated from 30 °C to 400 °C at a heating rate of 10 °C min^{-1} under nitrogen flow rate of 10 ml min^{-1}.

5. Results and discussion

5.1 Ac-impedance spectroscopy

The bulk ionic conductivity of polymer electrolytes is determined by using the equation as shown below.

$$\sigma = \frac{l}{R_b A}$$

where l is the thickness (cm), R_b is bulk resistance (Ω) and A is the known surface area (cm^2) of polymer electrolyte films. The semicircle fitting was accomplished to obtain R_b value. The R_b of the polymer electrolyte is calculated from the interception of high-frequency depressed semicircle with low-frequency spike.

Figure 5 depicts the logarithm of ionic conductivity with respect to PC mass fraction. As can be seen, the ionic conductivity increases with PC mass loadings, up to 10 wt% of PC. The optimum ionic conductivity of 5.10×10^{-7} Scm^{-1} is achieved with this mass fraction of PC. Plasticizing effect is the main attributor for this phenomenon. This effect would weaken the dipole–dipole interactions in the polymer chains and reduce the solvation of Li cations (Li^+) by polymer matrix. Hence, it promotes the ionic decoupling and enhances the dynamic free volume of the polymer system and thereby increases the ionic conductivity. It suggests that the plasticizer is not only weakening the polymer–polymer chain interactions, but also decreasing the dipole–ion interactions in the dopant salt. For the dipole–dipole interactions within the polymer chains, the hydrogen atom from the methyl group of PC may interact with the chloride anions in the polymer backbone. On the contrary, it proposes that the hydrogen from the methyl group of PC would weaken the O–Li bond of the $Li_2B_4O_7$. As a result, it promotes the dissociation of lithium cations from the bonding and hence favors the ionic transportation within the polymer matrix, improving the ionic conductivity.

In addition, the plasticizing effect lowers the T_g. Thus, it softens the polymer backbone and increases the segmental mobility when an electric field is applied onto the polymer electrolytes. Consequently, it disrupts the crystalline phase of polymer side chains and produces voids, which enables the easy flow of ions through polymer membrane when the electric field is applied. The ionic conductivity is eventually enhanced with these higher amorphous and more flexible polymer chains. The increase in ionic conductivity is also owing to the high dielectric constant of PC. High dielectric constant could allow the greater dissolution of $Li_2B_4O_7$ and offers a result in increasing the number of charge carriers, promoting the ionic hopping mechanism. The ionic conductivity is reduced to 4.20×10^{-8} Scm^{-1} with increasing the PC mass loadings further. This is suggestive of the decrease in effective number of charge carriers for ionic transportation as a result of the domain of short-range ion-plasticizer interactions within the polymer matrix (Stephan et al., 2002). Hence, the ionic conductivity is lower than of other plasticized-polymer electrolytes because of the reduced amount of lithium cations.

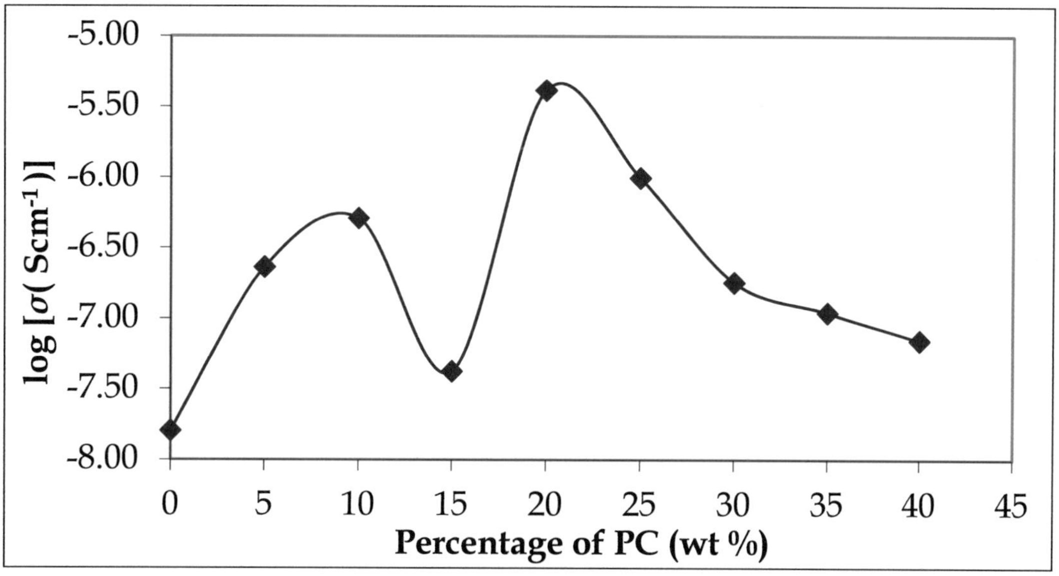

Fig. 5. The variation of logarithm of ionic conductivity of plasticized-based polymer electrolytes as a function of weight percentage of PC at ambient temperature.

Upon addition of 20 wt% of PC, the ionic conductivity is rising up further to the maximum level of 4.12×10^{-6} Scm^{-1} at room temperature. Again, the contribution from plasticizer is the main attributor for this enhancement of ionic conductivity. The incorporation of plasticizer increases the ionic conductivity through two ways. High plasticizer concentration would open up the narrow rivulets of plasticizer-rich region and lead to greater ionic migration. Moreover, it provides a large free volume of a relatively superior conducting region by reducing the crystalline degree of the polymer electrolytes (Rhoo et al., 1997; Stephan et al., 2000b). General expression of ionic conductivity of a homogenous polymer electrolyte is illustrated as below:

$$\sigma(T) = \sum_i n_i q_i \mu_i$$

where n_i is the number of charge carriers type of i, q_i is the charge of ions type of type of i, and μ_i is the mobility of ions type of i. Based on the equation above, the quantity and mobility of charge carriers are the main factors that could affect the ionic conductivity of polymer electrolytes as the charge of the mobile charge carriers are negligible. Therefore, it can be concluded that the mobility and concentration of mobile charge carriers have been optimized in SPC5 as it achieves the highest ionic conductivity compared to other polymer complexes. However, the ionic conductivity is drastically declined with increasing the PC concentration further. It is ascribed to the restricted ionic and segmental mobility of mobile charge carrier in a rigid polymer matrix (Cha et al., 2004).

5.2 Temperature dependence-ionic conductivity studies

The temperature dependence study of ionic conductivity is further investigated in order to understand the mechanism of ionic conduction in this plasticized–polymer electrolyte. SPC5 is chosen as it achieves the highest ionic conductivity. Figure 6 illustrates the logarithm of ionic conductivity against reciprocal absolute temperature of SPC5, from ambient temperature to 373 K. As expected, the ionic conductivity increases with temperature. Polymer expansion effect plays an important role in this phenomenon. The polymer matrix expands with temperature, which in turn to the formation of local empty spaces and voids for the segmental migration. Therefore, it facilitates the migration of ions and diminishes the ion clouds effect between the electrodes and electrolyte interface (Ramesh et al., 2010). The enhancement of charge carriers and segmental motions could assist the ionic transportation and compensate for the retarding effect of the ion cloud virtually, inducing to higher ionic conductivity.

A linear relationship is perceived in the figure with regression value of 0.99. Therefore, it can be concluded that SPC5 follows Arrhenius rules as its regression value is close to unity. In this thermally activated principle, the conductivity is expressed as below:

$$\sigma = A\,exp\,(\frac{-E_\mathrm{a}}{kT})$$

where A is the pre–exponential factor which is proportional to the amount of charge carriers, E_a is the activation energy, k is Boltzmann constant and T is the absolute temperature. The Arrhenius relationship indicates the presence of the hopping mechanism. This theory states the ion jumps from its normal position on the lattice to an adjacent equivalent but empty site. As the temperature increases, the vibrational modes of polymer segments are also increased. Thus, it weakens the interaction between the polar group of the polymer backbone and Li^+, and promotes the decoupling process of charge carriers from the segmental motion of polymer matrix, leading to formation of vacant sites in the polymer chain. Hence, the neighboring ions from adjacent sites tend to occupy these vacant sites and coordinate with the polymer chain again. Eventually, the ionic hopping mechanism is generated. In this study, it implies that the methyl group from PC and the C–H group from PVC could weaken O-Li coordinative bond of $Li_2B_4O_7$ through the hydrogen bonding. As a result, it initiates the decoupling of Li^+ from the bond and therefore generates the ionic hopping process. In order to probe the ion dynamic of polymer electrolytes further, activation energy (E_a) is determined by fitting it in Arrhenius equation as shown above. E_a is defined as the energy required to overcome the reorganization and reformation of the polymer chain with Li^+. Based on the calculation, the E_a of SPC5 is 0.08eV. This activation energy is considered low. Therefore, it can be concluded that Li^+ would break and re-bind the coordination bond easily with lower energy barrier.

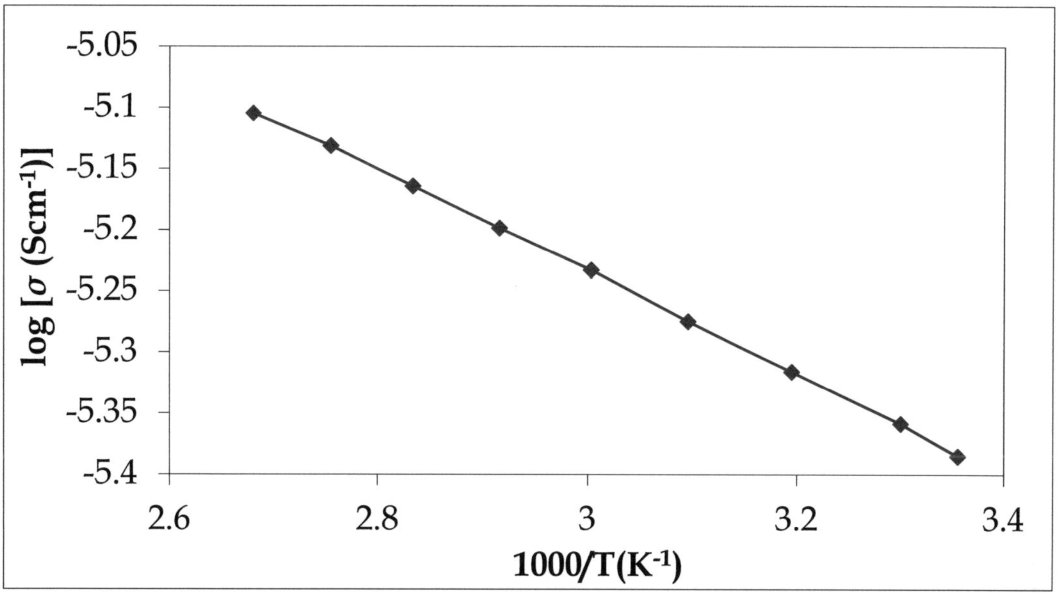

Fig. 6. Arrhenius plot of SPC5 in the temperature range of 298–373 K.

5.3 Fourier Transform Infrared (FTIR) studies

The FTIR spectra and description of vibration modes of pure PVC, pure $Li_2B_4O_7$, SPC1, PC and SPC5 are shown in Figures 7(a)–(e) and Table 2, respectively. Comparing SPC1 with pure PVC, there are 10 new peaks have been formed. All of these new peaks are the characteristic bonds of $Li_2B_4O_7$. Five new peaks have been detected in the wavenumber range of 1200 cm^{-1}–700 cm^{-1}. These peaks are assigned as B–O(B) stretching mode of BO_4 tetrahedral shape of $Li_2B_4O_7$ at 710 cm^{-1}, 815 cm^{-1}, 905 cm^{-1}, 1034 cm^{-1} and 1120 cm^{-1}. In contrast, for B–O(B) stretching mode of BO_3 triangle shape, $Li_2B_4O_7$ portrays two characteristic peaks at 1246 cm^{-1} and 1376 cm^{-1}. However, only one peak is observed at 1253 cm^{-1} for SPC1. This indicates the interaction between PVC and $Li_2B_4O_7$ and further reveals the decoupling of Li^+ from the B–O(B) coordinative bonds. Four new peaks at 451 cm^{-1}, 504 cm^{-1}, 565 cm^{-1} and 670 cm^{-1} are designated as O–B–O deformation mode of BO_4 tetrahedral in $Li_2B_4O_7$. All the vibration modes exhibit peak shifting, except the weak peak at 1331 cm^{-1} which is denoted as CH_2 deformation of PVC.

As shown in Figure 7(a), the transmittance peaks at 616 cm^{-1} and 969 cm^{-1} are corresponding to cis and trans C–H wagging modes, respectively. Upon addition of $Li_2B_4O_7$, these peaks are shifted towards higher wavenumber to 637 cm^{-1} and 973 cm^{-1}, respectively. Apart from that, they exhibit changes in shape. For cis wagging mode, it has been changed from weak peak to shoulder peak, whereas a medium peak has been changed to a broad band for trans wagging mode. A sharp peak is observed at 1067 cm^{-1} in Figure 7(a), which designated as C–H rocking mode of PVC. However, it turns to a broad band with inclusion of $Li_2B_4O_7$ and manifests a downward shift to 1062 cm^{-1}. The peak at 833 which corresponds to C–Cl stretching mode of PVC also shifted to 825 cm^{-1} upon impregnation of lithium salt. Similarly, C–H stretching mode of CH_2 group which is located at 1434 cm^{-1} for pure PVC spectrum also exhibits downward shift to 1420 cm^{-1}. This discloses the interactions between C–H, C–Cl and Li^+. It suggests that the Li^+ would be dissociated from O–Li interactive bond by forming hydrogen bonds to hydrogen atom from C–H group in PVC. Hence, these mobile Li^+ would re–interact with chloride anions in PVC as chloride anions have three electron lone pairs. The ionic hopping mechanism is eventually generated. The vibrational modes of characteristic peaks not only undergo the changes in shift and shape, but they also demonstrate the change in intensity. An apparent proof has been observed in the wavenumber region of 3000 cm^{-1}–2800 cm^{-1}. Two sharp peaks are located at 2867 cm^{-1} and 2979 cm^{-1}, and are denoted as CH_3 asymmetric stretching mode of PVC. The first transmittance peak is found to be shifted to higher wavenumber of 2910 cm^{-1}, meanwhile it has been moved to 2971 cm^{-1}, for latter peak. Upon the addition of $Li_2B_4O_7$, the intensity of both peaks is greatly reduced by comparing Figure 7(a) with 7(c). Regarding to the changes in peak intensity, changes in shape, changes in shift, formation of new peaks and disappearance of the peak, it reflects the establishment of polymer–salt complex.

In order to investigate the complexation between PC and polymer matrix, SPC5 is further examined as it achieves the maximum ionic conductivity. Comparing SPC1 with SPC5, seven new peaks have been formed. These peaks are denoted as ring deformation of PC, ring stretching and breathing modes of PC, C–O stretching mode of PC, B–O(B) stretching mode of BO_3 triangle shape of $Li_2B_4O_7$, C–H symmetric deformation mode of PC and C=O symmetric stretching mode of PC at 795 cm^{-1}, 959 cm^{-1}, 1054 cm^{-1} and 1187 cm^{-1}, 1351 cm^{-1}, 1388 cm^{-1} and 1794 cm^{-1}, respectively. Upon PC loadings, some of the characteristic peaks are disappeared. These peaks are the weak peaks in PC at 950 cm^{-1} and 910 cm^{-1}, and the shoulder peak of SPC1 at 1035 cm^{-1}. The characteristic peaks at 446 cm^{-1}, 503 cm^{-1}, 566 cm^{-1} and 668 cm^{-1} are designated as O–B–O deformation mode of BO_4 tetrahedral in $Li_2B_4O_7$. The weak peak at 446 cm^{-1} is originated from the medium sharp peak at 451 cm^{-1} in SPC1 spectrum. In term of intensity, this characteristic peak is reduced around 18%, from 23% to 5%, in transmittance mode. For the weak peaks at 503 cm^{-1} and 566 cm^{-1}, they show signs of changes in shape. It has been changed to broad band and slightly shifted from 504 cm^{-1} for the first peak. On the other hand, the latter peak displays a somewhat upward shift from 565 cm^{-1} and changed to shoulder peak. An oppose result is obtained for the peak at 702 cm^{-1}, which assigned as B–O(B) stretching mode of BO_4 tetrahedral in $Li_2B_4O_7$. This peak illustrates downward shift from 710 cm^{-1} to 702 cm^{-1} and the change in shape, from shoulder peak to weak peak.

Noticeable change in shape is observed in the wavenumber range of 700 cm^{-1}–600 cm^{-1}. A weak peak at 668 cm^{-1} with a shoulder peak at 637 cm^{-1} has been changed to two weak peaks at 670 cm^{-1} and 636 cm^{-1}, respectively, by doping PC into the polymer complex. As aforementioned, the peak at 670 cm^{-1} is the characteristic peak of O–B–O deformation mode of BO_4 tetrahedral, whereas cis C–H wagging mode in PVC is the assignment for the latter peak. Therefore, it implies the interaction between PVC, $Li_2B_4O_7$ and PC. There is another evidence to prove the complexation between PVC, $Li_2B_4O_7$ and PC at 830 cm^{-1}. Two shoulder peaks have been changed to a broad band. This arises from the combination of B–O(B) stretching mode of BO_4 tetrahedral shape of $Li_2B_4O_7$ at 815 cm^{-1}, C–Cl stretching mode of PVC at 825 cm^{-1} , and ring stretching and breathing modes of PC at 850 cm^{-1}. As explained in section 5.1, we propose that the hydrogen atom from methyl group in PC would break the O–Li coordination bond through hydrogen bonding. Thus, the dissociated Li^+ would interact with the chloride anion of the C–Cl interactive bond in PVC and ultimately form CH–Cl–Li linkage. A broad band is observed at 1062 cm^{-1} in SPC1 spectrum. However, this C–H rocking mode of PVC has been changed to weak shoulder peak at 1074 cm^{-1} with adulteration of PC. The effect of PC is also observed for the shoulder peak at 1120 cm^{-1}. As tabulated in Table 2, this peak is assigned as B–O(B) stretching mode of BO_4 tetrahedral of $Li_2B_4O_7$. Nevertheless, a medium sharp peak is attained at 1117 cm^{-1}. The change in shape is due to the merging of this stretching mode with C–O stretching mode of PC as a medium sharp peak is obtained at 1118 cm^{-1}, as shown in Figure 7(d). This interaction further proves the mechanism of complexation that we proposed as above.

SPC1 exemplifies two weak peaks at 1331 cm^{-1} and 1426 cm^{-1}. The first peak is designated as CH_2 deformation of PVC, whereas the C–H stretching mode of CH_2 group of PVC is for latter peak. Upon inclusion of PC, these two peaks are still appearing in the spectrum. The first peak is shifted upward to 1332 cm^{-1}, whereas the latter peak remains unchanged. Two more new weak peaks have been discovered in this band. These peaks are the B–O(B) stretching mode of BO_3 triangle of $Li_2B_4O_7$ and C–H symmetric deformation mode of PC at 1351 cm^{-1} and 1388 cm^{-1}, respectively. Moreover, the change in intensity is one of the aspects to determine the complexation of this plasticized-polymer electrolyte. The peak shifting of B–O(B) stretching mode of BO_3 triangle shape in $Li_2B_4O_7$ at 1253 cm^{-1} still remain the same. However, its peak intensity is slightly declined, from 28% to 21%, in transmittance mode, as illustrated in Figure 8. In contrast, the increase in peak intensity is obtained at high wavenumber range of 3000 cm^{-1}–2900 cm^{-1}. Only two peaks are observed in this range. Both of these peaks are denoted as CH_3 asymmetric stretching mode of PVC and shifted to 2912 cm^{-1} (from 2910 cm^{-1}) and 2975 cm^{-1} (from 2971 cm^{-1}). In term of intensity, the peaks are gradually increased. For the first peak, it rises up around 7%, from 5% to 12%, in transmittance mode. The peak intensity of latter peak enhances around 10%, from 7% to 17%, in transmittance mode. This reveals the interaction between PVC and PC and further verifies the establishment of polymer complex.

Some of the characteristic peaks of PC are not be found in the SPC5 spectrum compared to PC spectrum. These peaks include CH_2 bending deformational mode of CH_3 group of PC, in plane CH_2 scissoring mode of PC and the combinations of CH_2 rocking and ring breathing mode of PC at 1460 cm^{-1}, 1482 cm^{-1} and 1555 cm^{-1}, respectively. In PC spectrum, an intrinsic vibrational band of the C=O symmetric stretching mode is located at ~1800 cm^{-1}. This strong and broad band splits into two components (at 1787 cm^{-1} and 1900 cm^{-1}). The overtone is produced at 1900 cm^{-1} as a consequence of Fermi resonance of the C=O stretching mode with the ring breathing mode that lies at ~950 cm^{-1}. However, this overtone of PC is not being observed in SPC5 spectrum. This disappearance of characteristic divulges the interaction between PC and polymer system. The changes in position, changes in shape, changes in intensity, formation of new peaks and disappearance of peak infers the interaction between PVC, $Li_2B_4O_7$ and PC. Therefore, it can be concluded that PC is associated in the polymer matrix.

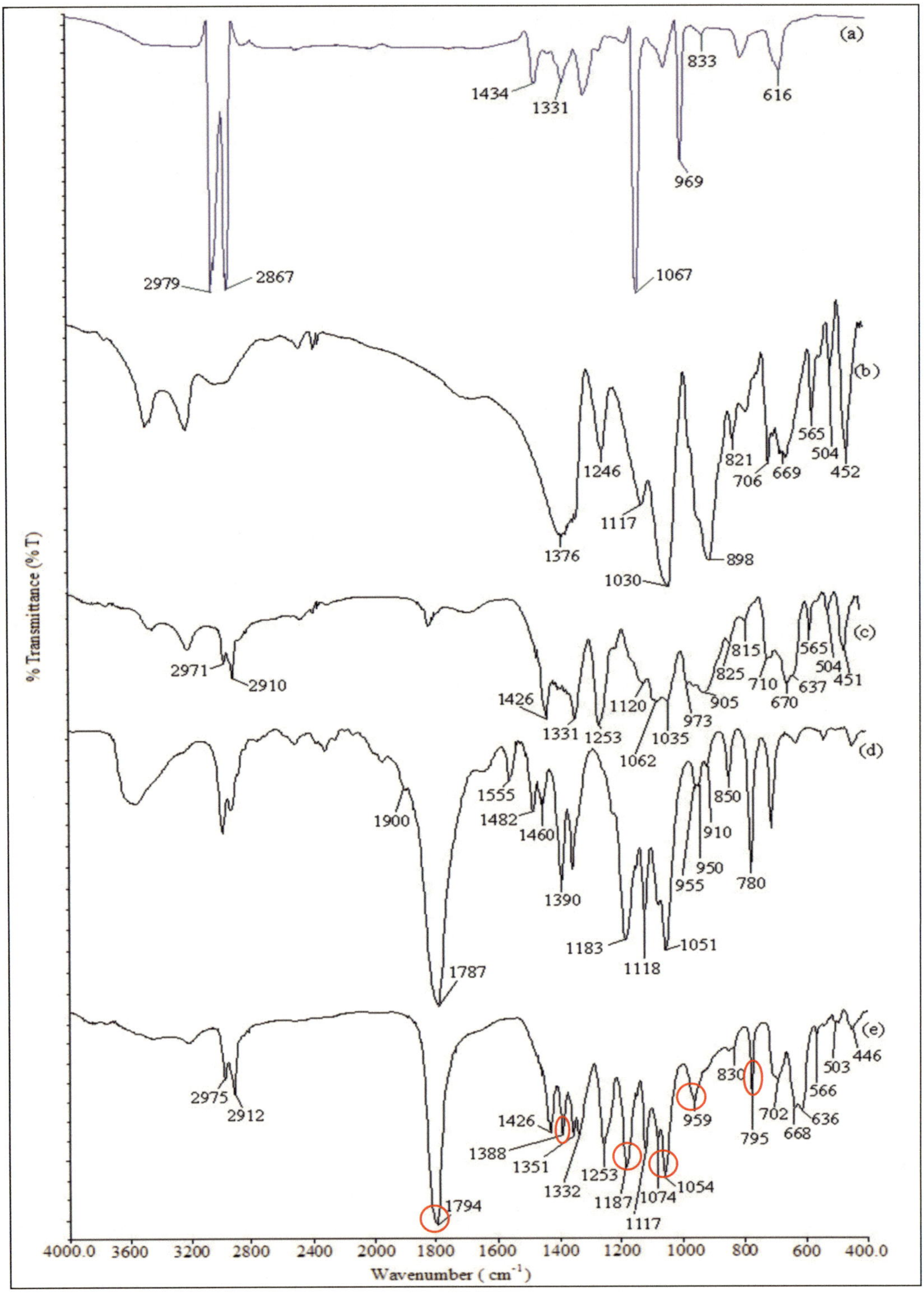

Fig. 7. The FTIR spectra of (a) pure PVC, (b) pure $Li_2B_4O_7$, (c) SPC1, (d) PC and (e) SPC5.

Description of vibration modes	Wavenumber (cm^{-1})			References
	PVC/ $Li_2B_4O_7$/PC	SPC1	SPC5	
O–B–O deformation mode of BO_4 tetrahedral of $Li_2B_4O_7$	452, 504, 565 and 669	451, 504, 565 and 670	446, 503, 566 and 668	Ge et al., 2007
Cis C–H wagging mode in PVC	616	637	636	Rajendran et al., 2008a
B–O(B) stretching mode of BO_4 tetrahedral shape of $Li_2B_4O_7$	706, 821, 898, 1030 and 1117	710, 815, 905, 1035 and 1120	702 and 1117	Ge et al., 2007
Ring deformation of PC	780	–	795	Deepa et al., 2004
C–Cl stretching mode of PVC	833	825	830	Li et al., 2006a
Ring stretching and breathing modes of PC	850, 910, 950 and 955	–	959	Deepa et al., 2004
Trans C–H wagging mode of PVC	969	973	Disappear	Achari et al., 2007
C–O stretching mode of PC	1051, 1118 and 1183		1054, 1117 and 1187	Deepa et al., 2004
C–H rocking mode of PVC	1067	1062	1074	Achari et al., 2007
B–O(B) stretching mode of BO_3 triangle shape of $Li_2B_4O_7$	1246 and 1376	1253	1253 and 1351	Ge et al., 2007
CH_2 deformation of PVC	1331	1331	1332	Rajendran et al., 2000b
C–H symmetric deformation mode of PC	1390	–	1388	Sharma and Sekhon, 2007
C–H stretching mode of CH_2 group of PVC	1434	1426	1426	Rajendran et al., 2008a
CH_2 bending deformation mode of CH_3 group of PC	1460	–	Not appear	Deepa et al., 2004
In plane CH_2 scissoring mode of PC	1482	–	Not appear	Deepa et al., 2004
Combinations of CH_2 rocking and ring breathing mode of PC	1555	–	Not appear	Deepa et al., 2004
C=O symmetric stretching mode of PC	1787	–	1794	Sharma and Sekhon, 2007
Overtone of PC (2×ring breathing mode of PC at 950 cm^{-1})	1900	–	Not appear	Sharma and Sekhon, 2007
CH_3 asymmetric stretching mode of PVC	2867 and 2979	2910 and 2971	2912 and 2975	Rajendran et al., 2000b

Table 2. Assignments of vibrational modes of pure PVC, pure $Li_2B_4O_7$, PC, SPC1 and SPC5.

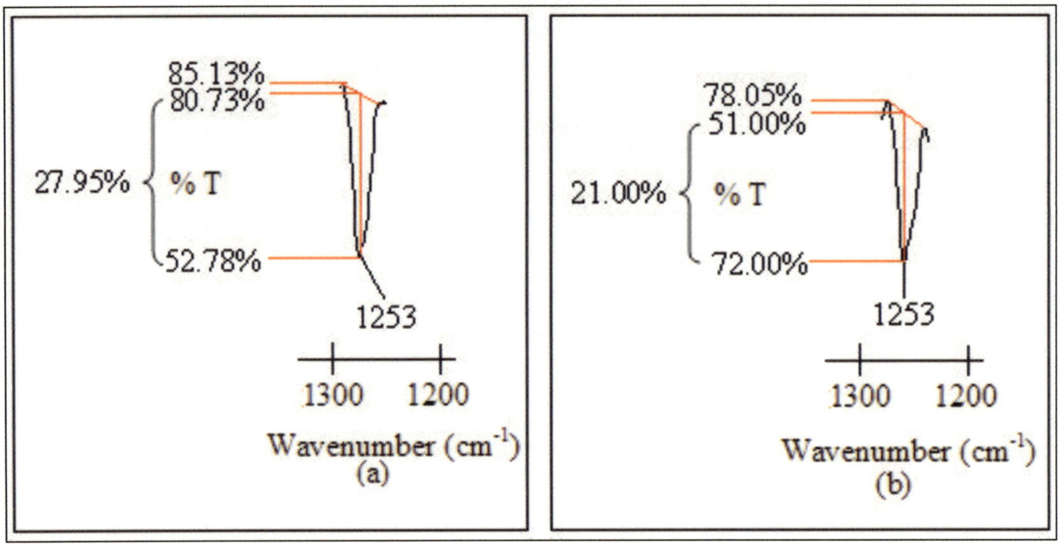

Fig. 8. The change in intensity of B–O(B) stretching mode of BO_3 triangle shape of $Li_2B_4O_7$ at 1253 cm^{-1} for (a) SPC1 and (b) SPC5.

5.4 Thermogravimetric analysis (TGA)

Figure 9 describes the thermogravimetric curves of SPC1, SPC5, SPC7 and SPC9. Two distinct stages have been observed in the temperature regime. The first weight loss is credited to the evaporation of residual THF solvent and dehydration of entrapped moisture (Ramesh et al., 2010). A moderate mass loss is initially observed. Pure PVC and SPC1 elucidate around 6% and 1% of mass losses at 159 °C and 150 °C, respectively. Even though the adulteration of PC increases the weight loss, but it also boosts up the decomposition temperature. As observed, the drop in weight is further increased with increasing the PC concentration. Around 11%, 31% and 32% of weight losses are attained for SPC5, SPC7 and SPC9 at 160 °C, 169 °C and 180 °C, respectively. After complete the dehydration, a stable weight is followed up in the thermal range.

Beyond this stable range, the weight of polymer complexes is drastically reduced in this stage. Dehydrochlorination process is the main contributor for this weight loss. At high temperature, the degrading products such as Cl free radicals are produced initially upon combustion. For further propagation, these free radicals would react with the methyl group of PC and hence break up the interactive bond, leading to the dehydrochlorination mechanism. The HCl cleavage would produce allyl chloride. Then, this allyl chloride favors the unzipping process and results in polyene linkage. This unzipping reaction induces many degradation reactions such as random chain scission reaction, depolymerization, inter-molecular transfer reaction and intra-molecular transfer reaction whereby dimers, trimers and oligomers are produced as well as polymer fragments. As a result, the monomer and oligomers which chemi-adsorbed onto the polymer matrix is volatilized in this region (Ramesh et al, 2011a). Pure PVC has mass loss of 63%, starting from 250 °C to 389 °C, with a residual mass of 31%. As can been seen, the weight losses have been improved by doping of $Li_2B_4O_7$ and PC. SPC1 delineates the mass loss of 46%, from 230 °C to 390 °C, with residual mass of around 51 %. The effect of PC onto the weight loss is further observed. SPC5 has lost around 45 % of its weight with around 43% residual mass, from 230 °C to 390 °C. In contrast, for SPC7, it is around 29% with 37% residual mass, starting from 238 °C to 388 °C. SPC9 starts to decompose at 239 °C and exemplifies a modest weight loss of 37% at 389 °C, with residual weight of 28%. Among the plasticized-polymer electrolytes, SPC5 portrays the lowest total weight loss of 57%. SPC5 is still a promising candidate as polymer electrolyte although its total weight loss is higher than SPC1. SPC5 exhibits excellent thermal stability as its stability is up to 230 °C, whereby the normal working range is 40-70 °C.

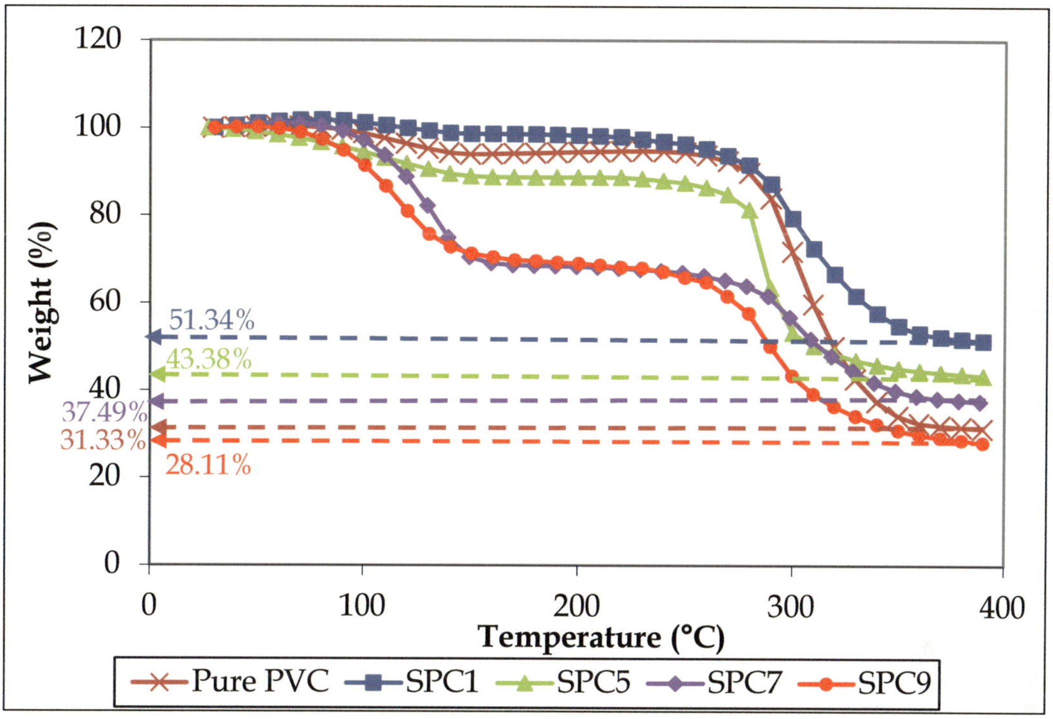

Fig. 9. Thermogravimetric analysis of pure PVC, SPC1, SPC5, SPC7 and SPC9.

6. Conclusion

The PVC-$Li_2B_4O_7$-PC plasticized-polymer system has been prepared and investigated in this project. Upon addition of 20wt% of PC (or designated as SPC5), the highest conductivity of 4.12×10^{-6} Scm^{-1} is achieved at ambient temperature. Plasticizer plays a fundamental role to weaken the interaction within the polymer matrix and hence increases the ionic conductivity with a flexible polymer backbone. The polymer electrolytes obey the Arrhenius behavior and indicate the ionic hopping mechanism, as proven in temperature dependence-ionic conductivity studies. In addition, FTIR studies help us to confirm the complexation of PVC-$Li_2B_4O_7$-PC system by determining the, changes in intensity, changes in shape and changes in shift, appearance and disappearance the peaks. Moreover, the thermal stability of the polymer films is contradictory to the PC mass loadings. By analyzing the TGA thermograms, it divulges that SPC5 exhibits good thermal stability in comparison with SPC7 and SPC9.

7. Further research

Upon addition of plasticizer, some limitations are obtained such as low flash point, slow evaporation, decreases in thermal, electrical and electrochemical stabilities. Low performances, for instance, small working voltage range, narrow electrochemical window, high vapor pressure and poor interfacial stability with lithium electrodes are the disadvantages of plasticized-gel polymer electrolytes (Pandey and Hashmi, 2009). Therefore, ionic liquid will be incorporated into the plasticizer in our further study. Room temperature ionic liquid (RTIL) possesses many attractive properties, such as wider electrochemical potential window (up to 6V), wider decomposition temperature range, non-toxicity and non-volatility as well as non-flammability with low melting point. Other features are better safety performance, relatively high ionic conductivity due to high ion content, and excellent thermal, chemical and electrochemical stabilities (Jiang et al., 2006; Cheng et al., 2007).

8. Acknowledgement

This work was supported by the Fundamental Research Grant Scheme (FRGS) from Ministry of Higher Education, Malaysia (FP009/2010B).

9. References

Achari, V.B., Reddy, T.J.R., Sharma, A.K. & Rao, V.V.R.N. (2007). Electrical, optical, and structural characterization of polymer blend (PVC/PMMA) electrolyte films. *Ionics*, Vol. 13, No. 5, (July 2007), pp. 349-354, ISSN 0947-7047.

Adebahr, J., Byrne, N., Forsyth, M., MacFarlane, D.R. & Jacobsson, P. (2003). Enhancement of ion dynamics in PMMA-based gels with addition of TiO_2 nano-particles. *Electrochimica Acta*, Vol. 48, No. 10, (June 2003), pp. 2099-2103, ISSN 0013-4686.

Ahmad, S., Ahmad, S. & Agnihotry S.A. (2005). Nanocomposite electrolytes with fumed silica in poly(methyl methacrylate): thermal, rheological and conductivity studies. *Journal of Power Sources*, Vol. 140, No. 1, (January 2005), pp. 151-156, ISSN 0378-7753.

Ahmad, S., Deepa, M. & Agnihotry, S.A. (2008). Effect of salts on the fumed silica-based composite polymer electrolytes. *Solar Energy Materials and Solar Cells*, Vol. 92, No. 2, (February 2008), pp. 184-189, ISSN 0927-0248.

Ali, A.M.M., Yahya, M.Z.A., Bahron, H., Subban, R.H.Y., Harun, M.K. and Atan, I. (2007). Impedance studies on plasticized PMMA-LiX [X: $CF_3SO_3^-$, $N(CF_3SO_2)^{2-}$] polymer electrolytes. *Materials Letters*, Vol. 61, No. 10, (April 2007), pp. 2026-2029, ISSN 0167-577X.

Barsoukov, E. & Macdonald, J.R. (2005). Fundamentals of Impedance Spectroscopy, In *Impedance Spectroscopy Theory, Experiment, and Applications.* , Evgenij Barsoukov and J. Ross Macdonald, pp. (1-26), John Wiley and Sons, ISBN 978-0-471-64749-2, United State: New Jersey.

Baskaran, R., Selvasekarapandian, S., Kuwata, N., Kawamura, J. & Hattori, T. (2007). Structure, thermal and transport properties of PVAc-$LiClO_4$ solid polymer electrolytes. *Journal of Physics and Chemistry of Solids*, Vol. 68, No. 3, (March 2007), pp. 407-412, ISSN 0022-3697.

Braun, D., Cherdron, H., Rehahn, M., Ritter, H. & Voit, B. (2005). Modification of Macromolecular Substances. *Polymer Synthesis: Theory and Practice.* (pp. 320-378),

Kunkel and Lopka, pp. (1-31), Springer, ISBN 978-3-540-20770-2, Berlin Heidelberg.

Byrappa, K. and Shekar, K.V. K. (1992). Hydrothermal synthesis and characterization of piezoelectric lithium tetraborate, $Li_2B_4O_7$ crystals. *Journal of Materials Chemistry,* Vol. 2, No. 1, (January 1992), pp. 13-18, ISSN 0959-9428.

Cha, E.H., Macfarlane, D.R., Forsyth, M. & Lee, C.W. (2004). Ionic conductivity study of polymer electrolytes containing lithium salt with plasticizer. *Electrochimica Acta,* Vol. 50, No. 2-3, (November 2004), pp. 335-338, ISSN 0013-4686.

Cheng, H., Zhu, C., Huang, B., Lu, M. & Yang, Y. (2007). Synthesis and electrochemical characterization of PEO-based polymer electrolytes with room temperature ionic liquids. *Electrochimica Acta,* Vol. 52, No. 19, (May 2007), pp. 5789-5794, ISSN 0013-4686.

Deepa, M., Sharma, N., Agnihotry, S.A., Singh, S., Lal, T. & Chandra, R. (2002). Conductivity and viscosity of liquid and gel electrolytes based on $LiClO_4$, $LiN(CF_3SO_2)_2$ and PMMA. *Solid State Ionics,* Vol. 152-153, (December 2002), pp. 253-258, ISSN 0167-2738.

Deepa, M., Agnihotry, S.A., Gupta, D. & Chandra, R. (2004). Ion-pairing effects and ion-solvent-polymer interactions in $LiN(CF_3SO_2)_2$-PC-PMMA electrolytes: a FTIR study. *Electrochimica Acta,* Vol. 49, No. 3, (January 2004), pp. 373-383, ISSN 0013-4686.

Endo, K. (2002). Synthesis and structures of poly(vinyl chloride). *Progress in Polymer Science,* Vol. 27, No. 10, (December 2002), pp. 2021-2054, ISSN 0079-6700.

Ganesan, S., Muthuraaman, B., Mathew, V., Madhavan, J., Maruthamuthu, P. & Suthanthiraraj, S.A. (2008). Performance of a new polymer electrolyte incorporated with diphenylamine in nanocrystalline dye-sensitized solar cell. *Solar Energy Materials and Solar Cells,* Vol. 92, No. 12, (December 2008), pp. 1718-1722, ISSN 0927-0248.

Ge, W., Zhang, H., Lin, Y., Hao, X., Xu, X., Wang, J., Li, H., Xu, H. & Jiang, M. (2007). Preparation of $Li_2B_4O_7$ thin films by chemical solution decomposition method. *Material Letters,* Vol. 61, No. 3, (February 2007), pp. 736-740, ISSN 0167-577X.

Gray, F.M. (1991). Polymer Electrolytes-Based Devices, In: Solid Polymer Electrolytes: Fundamentals and Technological Applications, Eishun Tsuchida, pp. (1-31), John Wiley & Sons, ISBN 0-85404-557-0, United Kingdom.

Gray, F.M. (1997a). What are Polymer Electrolytes?, In: *Polymer electrolytes,* Corron, J. A., pp. (1-26), The Royal Society of Chemistry, ISBN 978-0-471-18737-0, United Kingdom.

Gray, F.M. (1997b). What Materials are Suitable as Polymer Electrolytes?, In: *Polymer electrolytes,* Corron, J. A., pp. (31-78), The Royal Society of Chemistry, ISBN 978-0-471-18737-0, United Kingdom.

Gu, M., Zhang, Jun, Wang, X., Tao, H., Ge, L. (2006). Formation of poly(vinylidene fluoride) (PVDF) membranes via thermally induced phase separation. *Desalination,* Vol. 192, No. 1-3, (May 2006), pp. 160-167, ISSN 0011-9164.

Han, H.S., Kang, H.R., Kim, S.W. & Kim, H.T. (2002). Phase separated polymer electrolyte based on poly(vinyl chloride)/poly(ethyl methacrylate) blend. *Journal of Power Sources,* Vol. 112, No. 2, (November 2002), pp. 461-468, ISSN 0378-7753.

Jiang, J., Gao, D., Li, Z. & Su, G. (2006). Gel polymer electrolytes prepared by in situ polymerization of vinyl monomers in room-temperature ionic liquids. *Reactive and Functional Polymers,* Vol. 66, No. 10, (October 2006), pp. 1141-1148, ISSN 1381-5148.

Jorné, J. & Tobias, C.W. (1975). Electrodeposition of the alkali metals from propylene carbonate. *Journal of Applied Electrochemistry*, Vol. 5, No. 4, (Jan 1975), pp. 279-290, ISSN 0021-891X.

Kang, X. (2004). Nonaqueous Liquid Electrolytes for Lithium-Based Rechargeable Batteries. *Chemical Reviews*, Vol. 104, No. 10, (September 2004), pp.4303-4417, ISSN 0009-2665.

Kuila, T., Acharya, H., Srivastava, S.K., Samantaray, B.K. & Kureti, S. (2007). Enhancing the ionic conductivity of PEO based plasticized composite polymer electrolyte by $LaMnO_3$ nanofiller, *Materials Science and Engineering B*, Vol. 137, No. 1-3, (February 2007), pp. 217-224, ISSN 0921-5107.

Lee, C.C. & Wright, P.V. (1982). Morphology and ionic conductivity of complexes of sodium iodide and sodium thiocyanate with poly(ethylene oxide). *Polymer*, Vol. 23, No. 5, (May 1982), pp. 681-689, ISSN 0032-3861.

Li, W., Yuan, M. & Yang, M. (2006). Dual-phase polymer electrolyte with enhanced phase compatibility based on Poly(MMA-g-PVC)/PMMA. *European Polymer Journal*, Vol. 42, No. 6, (June 2006), pp. 1396-1402, ISSN 0014-3057.

Michael, M.S., Jacob, M.M.E., Prabaharan, S.R.S. & Radhakrishana, S. (1997). Enhanced lithium ion transport in PEO-based solid polymer electrolytes employing a novel class of plasticizers. *Solid State Ionics*, Vol. 98, No. 3-4, (June 1997), pp. 167-174, ISSN 0167-2738.

Nicotera, I., Coppola, L., Oliviero, C., Castriota, M. & Cazzanelli, E. (2006). Investigation of ionic conduction and mechanical properties of PMMA-PVdF blend-based polymer electrolytes. *Solid State Ionics*, Vol. 177, No. 5-6, (February 2006), pp. 581-588, ISSN 0167-2738.

Ning, W., Xiangxiang, Z., Haihui, L., Jianping, W. (2009). N,N-dimethylacteamide/lithium chloride plasticized starch as solid biopolymer electrolytes. *Carbohydrate Polymers*, Vol. 77, No. 3, (July 2009), pp. 607-611, ISSN 0144-8617.

Nishimoto, A., Watanabe, M., Ikeda, Y. & Kohjiya, S. (1998). High ionic conductivity of new polymer electrolytes based on high molecular weight polyether comb polymers. *Electrochimica Acta*, Vol.43, No. 10-11, (April 2008), pp. 1177-1184, ISSN 0013-4686.

Osinska, M., Walkowiak, M., Zalewska, A. & Jesionowski, T. (2009). Study of the role of ceramic filler in composite gel electrolytes based on microporous polymer membranes. *Journal of Membrane Science*, Vol. 326, No. 2, (January 2009), pp. 582-588, ISSN 0376-7388.

Pandey, G. P. & Hashmi, S. A. (2009). Experimental investigations of an ionic-liquid-based, magnesium ion conducting, polymer gel electrolyte. *Journal of Power Sources*, Vol. 187, No. 2, (February 2009), pp. 627-634, ISSN 0378-7753.

Pradhan, D. K., Samantaray, B. K., Choudhary, R. N. P. and Thakur, A. K. (2005). Effect of plasticizer on structure-property relationship in composite polymer electrolytes. *Journal of Power Sources*, Vol. 139, No. 1-2, (January 2011), pp. 384-393, ISSN 0378-7753.

Rajendran, S., Uma, T. & Mahalingam T. (2000a). Conductivity studies on PVC-PMMA-$LiAsF_6$-DBP polymer blend electrolyte. *European Polymer Journal*, Vol. 36, No. 12, (December 2000), pp.2617-2620, ISSN 0014-3057.

Rajendran, S. and Uma, T. (2000b). Experimental investigation on PVC-$LiAsF_6$-DBP polymer electrolyte systems. *Journal Power Sources*, Vol. 87, No. 1-2, (April 2000), pp. 218-222, ISSN 0378-7753.

Rajendran, S., Mahendran, O. & Kannan, R. (2002). Characterisation of [(1- x) PMMA- x PVdF] polymer blend electrolyte with Li^+ ion. *Fuel*, Vol. 81, No. 8, (May 2002), pp. 1077-1081, ISSN 0016-2361.

Rajendran, S., Sivakumar, M. & Subadevi, R. (2004). Investigations on the effect of the various plasticizers in PVA-PMMA solid polymer blend electrolytes. *Materials Letter*, Vol. 58, No. 5, (February 2004), pp. 641-649, ISSN 0167-577X.

Rajendran, S., Prabhu, M.R. & Rani, M.U. (2008a). Ionic conduction in poly(vinyl chloride)/poly(ethyl methacrylate)-based polymer blend electrolytes complexed with different lithium salts. *Journal of Power Sources*, Vol. 180, No. 2, (June 2008), pp. 880-883, ISSN 0378-7753.

Rajendran, S. & Sivakumar, P. (2008b). An investigation of PVdF/PVC-based blend electrolytes with EC/PC as plasticizers in lithium battery applications. *Physica B:Condensed Matter*, Vol. 403, No. 4, (March 2008), pp. 509-516, ISSN 0921-4526.

Ramesh, S. & Arof, A.K. (2000). Electrical conductivity studies of poly(vinylchloride) based electrolytes with double salt system. *Solid State Ionics*, Vol. 136-137, (November 2000), pp. 1197-1200, ISSN 0167-2738.

Ramesh, S. & Arof, A.K. (2001). Ionic conductivity studies of plasticized poly(vinyl chloride) polymer electrolytes. *Materials Science and Engeneering B*, Vol. 85, No. 1, (August 2001), pp. 11-15, ISSN 0921-5970.

Ramesh, S. & Chai, M.F. (2007). Conductivity, dielectric behavior and FTIR studies of high molecular weight poly(vinyl chloride)-lithium triflate polymer electrolytes. *Materials Science and Engineering B*, Vol. 139, No. 2-3, (December 2007), pp. 240-245, ISSN 0921-5107.

Ramesh. S. & Ng, K.Y. (2009). Characterization of polymer electrolytes based on high molecular weight PVC and Li_2SO_4, *Current Applied Physics*, Vol. 9, No. 2, (March 2009), pp. 329-332, ISSN 1567-1739.

Ramesh, S., Chiam-Wen, Liew, Morris, E. & Durairaj, R. (2010). Effect of PVC on ionic conductivity, crystallographic structural, morphological and thermal characterizations in PMMA-PVC blend-based polymer electrolytes. *Thermochimica Acta*, Vol. 511, No. 1-2, (August 2010), pp. 140-146, ISSN 0040-6031.

Ramesh, S., Chiam-Wen, Liew & Ramesh, K. (2011a). Evaluation and investigation on the effect of ionic liquid onto PMMA-PVC gel polymer blend electrolytes. *Journal of Non-Crystalline Solids*, Vol. 357, No. 10, (March 2011), pp. 2132-2138, ISSN 0022-3093.

Ramesh, S. & Chao, L.Z. (2011b). Investigation of dibutyl phthalate as plasticizer on poly(methyl methacrylate)-lithium tetraborate based polymer electrolytes. *Ionics*, Vol. 17, No. 1, (September 2011), pp. 29-34, ISSN 0947-7047.

Rhoo, H.J., Kim, H.T., Park, J.K. & Hwang, T.S. (1997). Ionic conduction in plasticized PVC/PMMA blend polymer electrolytes. *Electrochimica Acta*, Vol. 42, No. 10, (January 1997), pp. 1571-1579, ISSN 0013-4686.

Selvasekarapandian, S., Baskaran, R., Kamishima, O., Kawamura, J. & Hattori, T. (2006). Laser Raman and FTIR studies on Li^+ interaction in PVAc-$LiClO_4$ polymer electrolytes. *Spectrochimica Acta Part A: Molecular and Biomolecular Spectroscopy*, Vol. 65, No. 5, (December 2006), pp. 1234-1240, ISSN 1386-1425.

Sharma, J. P. & Sekhon, S.S. (2007). Nanodispersed polymer gel electrolytes: Conductivity modification with the addition of PMMA and fumed silica. *Solid State Ionics*, Vol. 178, No. 5-6, (March 2007), pp. 439-445, ISSN 0167-2738.

Sivakumar, M., Subadevi, R., Rajendran, S., Wu, N.L. & Lee, J.Y. (2006). Electrochemical studies on [(1−*x*)PVA-*x*PMMA] solid polymer blend electrolytes complexed with $LiBF_4$, *Materials Chemistry and Physics*, Vol. 97, , No. 2-3, (June 2006), pp. 330-336, ISSN 0254-0584.

Stephan, A.M., Kumar, T.P., Renganathan, N.G., Pitchumani, S., Thirunakaran, R. & Muniyandi, N. (2000a). Ionic conductivity and FT-IR studies on plasticized PVC/PMMA blend polymer electrolytes. *Journal of Power Sources*, Vol. 89, No. 1, (July 2000), pp. 80-87, ISSN 0378-7753.

Stephan, A. M., Renganathan, N.G., Kumar, T.P., Thirunakaran, R., Pitchumani, S., Shrisudersan, J. & Muniyandi, N. (2000b). Ionic conductivity studies on plasticized PVC/PMMA blend polymer electrolyte containing $LiBF_4$ and $LiCF_3SO_3$. *Solid State Ionics*, Vol. 130, No. 1-2, (May 2000), pp. 123-132, ISSN 0167-2738.

Stephan, A.M., Saito, Y., Muniyandi, N., Renganathan, N.G., Kalyanasundaram, S. & Elizabeth, R.N. (2002). Preparation and characterization of PVC/PMMA blend polymer electrolytes complexed with $LiN(CF^3SO^2)^2$. *Solid State Ionics*, Vol. 148, No. 3-4, (June 2002), pp. 467-473, ISSN 0167-2738.

Stephan, A.M. (2006). Review on gel polymer electrolytes for lithium batteries. *European Polymer Journal*, Vol. 42, No. 1, (January 2006), pp. 21-42, ISSN 0014-3057.

Suthanthiraraj, S.A., Sheeba, D.J. & Paul, B.J. (2009). Impact of ethylene carbonate on ion transport characteristics of PVdF-$AgCF_3SO_3$ polymer electrolyte system. *Materials Research Bulletin*, Vol. 44, No. 7, (July 2009), pp. 1534-1539, ISSN 0025-5408.

Tobishima, S.I. & Yamaji, A. (1984). Ethylene carbonate-propylene carbonate mixed electrolytes for lithium batteries. *Electrochimica Acta*, Vol. 29, No. 2, (February 1984), pp. 267-271, ISSN 0013-4686.

Tobishima, S.I., Hayashi, K., Nemoto, Y., Yamaki, J.-I. (1997). Ethylene carbonate/propylene carbonate/2-methyl-tetrahydrofuran ternary mixed solvent electrolyte for rechargeable lithium/amorphous V_2O_5-P_2O_5 cells. *Electrochimica Acta*, Vol. 42, No. 11, (May 1997), pp. 1709-1716, ISSN 0013-4686.

Xu, J., Lu, B., & Fan, S. (2004). Industrial Growth of Lithium Tetraborate ($Li_2B_40_7$) Piezocrystal and Its SAW Applications. *Proceedings of 7th international conference on Solid-State and Integrated Circuits Technology*, ISBN 0-7803-8511-X, China:Beijing, October 2004.

Xu, J.J. & Ye, H. (2005). Polymer gel electrolytes based on oligomeric polyether/cross-linked PMMA blends prepared via in situ polymerization. *Electrochemistry Communications*, Vol. 7, No. 8, (August 2005), pp. 829-835, ISSN 1388-2481.

Yang, Y., Zhou, C.H., Xu, S., Hu, H., Chen, B. L., Zhang, J., Wu, S.J., Liu, W., Zhao, X. Z., et al. (2008). Improved stability of quasi-solid-state dye sensitized solar cell based on poly(ethylene oxide)-poly(vinylidene fluoride) polymer-blend electrolytes. *Journal of Power Sources*, Vol. 185, No. 2, (December 2008), pp. 1492-1498, ISSN 0378-7753.

Zhang, P., Yang, L.C., Li, L.L., Ding, M.L., Wua, Y.P. & Holze, R. (2011). Enhanced electrochemical and mechanical properties of P(VDF-HFP)-based composite polymer electrolytes with SiO_2 nanowires. *Journal of Membrane Science*, Vol. 379, No. 1-2, (September 2011), pp. 80-85, ISSN 0376-7388.

10

Health Risk Assessment of Plasticizer in Wastewater Effluents and Receiving Freshwater Systems

Olalekan Fatoki[1], Olanrewaju Olujimi[1,*], James Odendaal[1] and Bettina Genthe[2]
[1]*Faculty of Applied Sciences*
Cape Peninsula University of Technology, Cape Town,
[2]*CSIR, Stellenbosch,*
South Africa

1. Introduction

A variety of human activities e.g. agricultural activities, urban and industrial development, mining and recreation, significantly alter the quality of natural waters, and changes the water use potential (Spinks et al., 2006; Madungwe and Sakuringwa, 2007). The key to sustainable water resources is, therefore to ensure that the quality of water resources are suitable for their intended uses, while at the same time allowing them to be used and developed to a certain extent. Water quality management, therefore involves the maintenance of the fitness for use of water resources on a sustained basis, by achieving a balance between socio-economic development and environmental protection. Approximately 40 000 small-scale farmers, 15 000 medium-to-large-scale farmers, 120 000 permanent workers, and an unknown number of seasonal workers are involved in irrigation farming, which consumes approximately 51 to 61 % of South Africa's water on some 1,3 million hectares (Backeberg, 1996; Blignaut and Heerden, 2008). Irrigation farming contributes 25 to 30 % of South Africa's agricultural output. Agriculture is crucially important to the basic food security of the poor, who constitute 40 % of the population of 42 million, and who are overwhelmingly concentrated in rural areas and (peri-) urban townships (Blignaut and Heerden, 2008).

Like many countries in the world, water scarcity is becoming a major problem in South Africa (Marcucci & Tognotti, 2002; Oweis & Hachum, 2009; Komnenic *et al.*, 2009) as dams serving communities with drinking water and water for daily household use, have been less than 30% full in recent years (Qiao *et al.*, 2009; Malley *et al.*, 2009). River water, in combination with groundwater, effluents from wastewater treatment plants, is considered a suitable alternative as a utilisable and potable water source (Blignaut and Heerden, 2008). To complement scare water resources, there has been increase in the number of wastewater facilities in many countries. This is to forestall the outbreak of environmental pollution and spread of diseases, remove conventional pollutants (such as ammonia and phosphate), and to maintain and restore the biologic integrity of surface waters (Wang *et al.*, 2005; Sun *et al.*, 2008). Domestic and industrial wastewaters are significant sources of endocrine disrupting chemicals

(EDCs) to the receiving surface, coastal waters and regional environments (Ahel *et al.*, 1994; Ahel *et al.*, 1996; Ying *et al.*, 2002; Vethaak *et al.*, 2005; Voutsa *et al.*, 2006; Zuccato *et al.* 2006).

South African rivers are steadily becoming more contaminated and in some cases even toxic, due to urbanization, industrialization and malfunctioning of wastewater treatment plants in the cities (Fatoki et al., 2004; Jackson et al., 2007; Jackson et al., 2009). Water quality in South Africa has been a major debate considering the water consumption trend in the country both for agricultural development, recreational purposes and domestication usage. Of the major water pollutants that have been relegated to the background in South Africa is Phthalate esters (PE). There was no local interim guidelines for PE in freshwater systems in South Africa, thus pollution of freshwater systems through industrial activities could not be punished. However, water quality is of paramount importance in this country.

Phthalate ester is synthetic compound commonly used as a plasticizer to impart flexibility, workability, and durability to polymers such as polyvinyl chloride. Also, this compound is used in a wide variety of products such as paints, adhesives, inks and cosmetics (Ling et al., 2007; Huang et al., 2007). As a result, PE has become ubiquitously distributed in the environment and easily finds their ways into the river systems through both dry and wet deposition (Yuan et al., 2002; Yuan et al., 2008). PE is considered to be a potential carcinogen, teratogen, and mutagen. Their toxicity to human beings and aquatic organisms is of deep concern (Mylchreest et al., 1999; Awal et al., 2004; Fatoki et al., 2010).

Furthermore, PE acts as endocrine disruptors, which could alter reproductive functions and exert distinct effects on male reproductive organs due to anti-androgenic effects (Latini et al., 2006; Lambrot et al., 2009; Vo et al., 2009). The aim of this study was to assess the potential human impacts health associated with PE found in the final effluent from wastewater treatment plants and river water receiving effluent wastes.

1.1 The risk assessment framework

In recent decades, the interest about environmental issues has increased very quickly. Not only to the natural scientists, but other active members of the society (politicians, industrialists and the general public), have paid much attention in all aspects related to the environment, in general, and environment protection, in particular. In this context, environmental pollution has been one of the fields where more efforts have been aimed to control. Because of the lack of environmental consciousness and technical capacity, many industries released toxic substances into the air, water and soil, for a number of years. As a first consequence, levels of pollution in areas surrounding industrial sites became much higher than background (unpolluted) zones. Recently, implementation of legislative measures carried out by public administrations has obliged to companies to improve their production processes in order to reduce the pollutant emissions.

The concern resulting from the potential exposure to contaminants initiated the development of methodologies that evaluate the consequences that those contaminants can have on environment and human health. Among these methods, risk assessment has been one of the most widely used. Risk assessment is a formalized process for estimating the magnitude, likelihood, and uncertainty of environmentally induced health effects (Sexton *et al.*, 1995). In 1983, the US National Research Council (NRC), in the so-called "Red Book", defined a series of principles to be considered for human health risk assessment, and defined it as a process in which information is analyzed to determine if an environmental hazard might cause harm to exposed persons and ecosystems (NRC, 1983).

In addition to definition, NRC proposed a framework for human health risk assessment, which involved 4 basic steps (NRC, 1993). The four steps of the process are:

1. Hazard identification
2. Dose-response assessment
3. Exposure assessment and
4. Risk characterization.

1.1.1 Hazard identification

This step can be defined as the qualitative determination of whether or not a particular hazardous agent is associated with health effects of sufficient importance to warrant further scientific investigations. Different kinds of tools (QSAR, short-term toxicity test) are used in order to estimate the chemical damage of a single substance. When establishing the hazard from industrial sources, the chemicals are also identified according to measurements of amount and typology of emissions.

1.1.2 Dose-response assessment

This component is focused on examining quantitative relationships between the magnitude of the exposure (or dose) and the probability of occurrence of adverse effects in the population. Usually, dose-response assessment is based on extrapolations from data about laboratory animals, which have been given high-doses of toxicant and monitored accordingly.

1.1.3 Exposure assessment

Exposure assessment may be defined as the quantitative determination of the extent of exposure of the population to the hazardous agent in question. Since they provide a real knowledge of the state of pollution of an area, data obtained in the environmental monitoring are commonly used as a starting point. Factors that need to be considered include frequency and duration of exposure, rates of uptake or contact, and rate of absorption (NRC, 1993). Other factors in assessing exposure include release patterns, cumulative versus non-cumulative exposure, persistence, failure of exposure controls, quality of data and quality of models.

1.1.4 Risk characterization

This fourth component can be defined as the description of the nature and magnitude of the risk, expressed in terms which are comprehensible to decision makers and the public. Information acquired in the previous 3 steps is integrated in order to communicate the overall meaning of, and confidence in, the hazard, exposure, and risk conclusions. Risk is expressed as a probability of suffering a particular kind of harm from a hazard to a specified group of population (Bennion *et al.*, 2005). Moreover, qualitative and quantitative uncertainty related to risk must be also supplied.

The project aimed at determining the potential health risks that may be associated with using river water and treated effluent from wastewater treatment plants in Cape Town. Since phenols and phthalate esters were placed on the United State Environmental Protection Agency list as priority pollutants, both phenols and phthalate esters congeners were analyzed in water samples. However, emphasis is more on the phthalate esters congeners. The derivatization of the phenolic congeners did not in any way affect the intensity of the phthalate esters congeners included in this study (Olujimi et al., 2011b).

2. Materials and method

2.1 Study areas

Influents and effluents from six wastewater treatment plants namely; Athlone, Bellville (which consist of the Old and New plants), Kraaifontein, Potsdam, Stellenbosch and Zandvliet) were investigated for the occurrence of seventeen organic compounds (eleven priority phenols and six phthalate esters). Five of these wastewater treatment plants (WWTPs) were located in the City of Cape Town, while one is located in Stellenbosch. Rivers associated with each treatment plant are: Athlone - Vygekraal River; Bellville - Kuils River; Kraaifontein -Mosselbank River; Potsdam - Diep River; Zandvliet - Kuils River and Stellenbosch -Veldwachters River. Five of the WWTPs and associated rivers investigated are presented in Figure 1. Samples were taken at the point of discharge, as well as upstream and downstream from point of discharge (about 1-2km) to evaluate the possible impact of effluent on organic compounds load on the aquatic environment. The geographical location, population equivalent and treatment processes of the investigated treatment plants are presented in Table 1.

WWTPID	Geographical Location of plant	People equivalent	Source	Treament Process	River
A	S33.5709° E18.3048°	900,000	Domestic Industrial	S + G + Sed + AS (BNR) + Sed + Chl + AD + Dew	Vygekraal River
B	S33.5923° E18.4332°	591,000	Domestic Industrial	S + G + EAAS (N) + Sed + UVdis + Dew	Kuils River 1
C	S33.82539° E18.70442°	133,000	Domestic	S + G + Sed + AS (N) + Sed + Chl + AD + Dew	Mosselbank River
D	S33.5070° E18.3108°	385,000	Domestic Industrial	S + G + Sed + AS (N) + Sed + Chl + AD + Dew	Diep River
E	S33.94345° E18.82492°	N/K	Domestic Industrial	S + G + Sed + FB + AS (BNR) + Sed + Chl + AD + Dew	Veldwatchers River
F	S34.0312° E18.4259°	400,000	Domestic Industrial	S + G + Sed + AS (N) + Sed + Chl + AD + Dew	Kuils River 2

Abbreviations: S = Screening; G = Grit removal; Sed = Sedimentation; AS = Activated Sludge; EAAS = Extended Aeration Activated Sludge; N = Nitrogen; BNR = Biological nutrient removal; Chl = Chlorination; UVdis = UV disinfection; AD = Anaerobic digestion; FB = Filter bed; N/K = Not known; WWTP ID = Wastewater treatment plant identification.

Table 1. Description of the six wastewater treatment plants investigated.

Fig. 1. Map showing five of the five of the wastewater treatment plants.

2.2 Chemicals and reagents

Analytical grade phenol (PH) 99.9 %, 2-nitrophenol (2-NP) 99 %, 4-nitrophenol (4-NP) 99 %, 2,4-dinitrophenol (2,4-DNP) 99.7 %, 4,6-dinitro-2-methylphenol (DNMP) 98 %, 2,4-dimethylphenol (2,4-DMP) 98 %, 2-chlorophenol (2-CP) 99.8 %, 4-chlorophenol (4-CP) 99 %,

2,4-dichlorophenol (2,4-DCP) 100 %, 4-chloro-3-methylphenol (4-C-3MP) 99 %, pentachlorophenol (PCP) 99.6 %, dimethyl phthalate (DMP), diethyl phthalate (DEP) 99 %, benzybutyl phthalate (BBP) 98 %, dioctyl phthalate (DOP) 99 %, diethylhexylphthalate (DEHP) 99 %, dibutyl phthalate (DBP) 99 % were purchased from Superlco (Bellefonte, PA USA). Helium (99.999 %) is supplied by Afrox gas, South Africa, Potassium Carbonate, acetic anhydride were supplied by Separations (South Africa). The solvents (methanol, n-hexane, acetone and acetonitrile) were of analytical grade from Sigma Aldrich and were further purified by distillation. Separate stock solutions (1000 mgl^{-1}) of individual congeners were prepared in methanol. A working mixture containing each compound at 10 mgl^{-1} was also prepared and stored at 4°C in the dark. Milli-Q water used was from apparatus Millipore (Bedford, MA, USA).

2.3 Derivatization procedure

Some EDCs such as phenols with hydroxyl group within the molecule have to be derivatized with N-Methyl-N- (Tert-Butyldimethylsilyl) trifluoroacetamide (MTBSTFA), which results in the formation of tert-butyltrimethylsilyl (TBMS) derivatives. The high polarity of the phenolic compounds gives rise to poor chromatographic performance and as a consequence derivatization was carried out. The phenol-silylate is more volatile and affords better detection limits when using gas chromatography (GC). The standard mixture was derivatized according to Olujimi *et al.* (2011b). Briefly, 1 ml of the standard mixture (phenols and phthalate esters) was measured into sample vial and blown to dryness under gentle flow of nitrogen gas. The dried standard mixture was reconstituted with 50 μl acetonitrile and 50 μl silylating reagent N-Methyl-N- (Tert-Butyldimethylsilyl) trifluoroacetamide (MTBSTFA) and mixed in a vortex for 90 s. The solution was derivatized at 90 °C for 20 min in a GC oven. The sample was cooled down to room temperature and 1 μl was injected into the GC-MS for analysis. The stepwise derivatization procedure is shown in Figure 2. The GC-MS parameters used for the analysis is presented in Table 2 after initial optimization studies.

2.4 Determination of limits of detection and quantification GC-MS

Lower concentration standards were prepared through serial dilution of individual standard of phenols and phthalate esters as well as the mixture standards. 1 μl aliquots of each of the standard was injected into GC, to determine the lowest concentration. Different procedures for the determination of limits of detections (LODs) and limit of quantifications (LOQs) are reported in the literature. These limits can be experimentally estimated from the injection of serially diluted standard solutions or extracts of fortified water samples until the signal-to-noise ratio (s/n) ratio reaches a value of three. LOD was estimated as three times the noise level of the baseline in the chromatogram, while the limit of quantification (LOQ) is set at three times the LOD. For this study, LOD and LOQ were calculated using the equations below:

$$LOD = 3.3 \times Sb/a \tag{1}$$

and

$$LOQ = 10 \times Sb/a \tag{2}$$

where a is the slope and Sb is the standard deviation of the y-intercept (De Sousa et al., 2003).

2.5 Solid phase extraction (SPE) for water samples

C18-E cartridges (strata, 500 mg/ 6 ml) from Separations Limited were used for the extraction of phenols and phthalates from water samples based on recoveries obtained for phenols using HPLC (Olujimi et al., 2011a). Prior to the sample processing, the cartridges were fitted onto a vacuum manifold (Supelco) connected to pump. The cartridges were conditioned with 5 ml of n-hexane:acetone (50:50, v/v), followed sequentially by 5 ml of methanol and 10 ml of Milli-Q purified water (purified by Milli-Q System, Millipore, Bedfore, MA, USA). Prior to extraction of each 500 ml, water samples were filtered on vacuum using a 0.22 µm filter to remove suspended particulate matter that might block the SPE cartridges. Hydrochloric acid (37 %) was used to adjust the pH of the water sample to pH ≤ 3 before passing it through the conditioned cartridge. The cartridge were then rinsed with 5 ml of Milli-Q water and left on the vacuum manifold for 30 min to dry (-70 Kpa). The retained analytes of interest were eluted with 3.5 mL of methanol followed by 3.5 ml of n-hexane:acetone (50:50, v/v) into 10 ml glass vials. This was blown to dryness on hot plate at 70 °C under gentle flow of nitrogen gas. The retained analytes were then derivatized according to the procedure described in section 2.3 (Figure 2).

1 ml of standard mixture/ extracted analytes

↓

Blow to dryness on hot plate at 70 °C
under gentle flow of nitrogen

↓

50 µl each of Acetonitrile and MTBSTFA

↓

Vortex mix (90 s)

↓

Derivatized at 90 °C for 20mins in GC oven
cool to room temperature

↓

Analyze on GC-MS

Fig. 2. Derivatization procedure for silylation.

Gas chromatography		Mass spectrometer	
GC-MS	Agillent 6890N	5975	
Capillary Column	DB-5MS, 30 m x 0.25 mm i.d. (0.25 μm film thickness)	Ion source	Electron impact ionization, 70 ev
Carrier gas	Helium, purity: 99.999 %	Ion source temperature	230 ^{o}C
Injector parameters	1 μl splitless, injection temperature 260 ^{o}C	Inlet temperature	260 ^{o}C
Oven temperature	80 ^{o}C (1 min) -5 ^{o}C min^{-1} 150 ^{o}C held for 1 min, then to 280 ^{o}C at 12 ^{o}C (7 min), carrier gas flow rate: 1.0 mlmin^{-1} Post run temperature: 300 ^{o}C (2 min)	Transfer line Scan mode (m/Z)	280 ^{o}C 50-450

Table 2. Gas Chromatography and Mass Spectrometer Parameters.

2.6 Seasonal sampling protocol

Water samples for organic compounds analysis were collected from the wastewater treatment plants and rivers on quarterly basis. This was to observe the possible impact of seasonal variation on organic compounds in wastewater treatment plants and possible impact this could have on the concentration of congeners in the freshwater systems. Sampling started in April 2010 and ended in March 2011.

2.7 Quality assurance and quality control (QA/QC) for GC-MS

Spiked procedural blanks, solvent blanks and control samples were included in each batch of analyses. Blanks and controls were treated similarly as the samples and analyzed after every sample injection. A calibration standard solution of 50 μgl^{-1} was injected in duplicate to monitor the instrumental sensitivity and reproducibility every time before sample analyses.

2.8 Health risk assessment

A human health risk assessment was conducted to provide an indication of whether the organic compounds or heavy metals detected in the water samples tested may cause adverse health effects to human. The methodology used to assess this potential human health risk was that described by US-EPA (1988, 1996) and the WHO (2002). The exposures considered in the assessment include:

a. Ingestion through drinking of final effluents or river water,
b. Dermal absorption due to daily washing/bathing in the river water,
c. Irrigating farm lands with final effluent or river water,
d. If fish from these areas is consumed.

Human exposure to toxic effects are expressed in terms of average daily dose (ADD) which is the amount of substance taken into the body on daily basis during the exposure period calculated

$$ADD = (C_{medium} \ x \ IR \ x \ ED \ x \ F_c) / BW \ x \ AT \ (\text{mg/kg·d}) \quad (5)$$

where:

ADD = average daily dose C_{medium} = concentration in the contaminated water
IR = daily intake rate
ED = exposure duration
$F_{c,}$ = the fraction contaminated
BW = body weight
AT = lifetime averaging time

For risk of carcinogens for exposures that last less than lifetime, the dose is adjusted using the formula:

$$LADD = ADD \text{ x } (ED/Lft) \quad (6)$$

where: Lft is lifetime

2.8.1 Non-cancer toxic effects (Hazard Quotient)

For agents that cause non-cancer effects, a Hazard Quotient (H.Q) was calculated, comparing the expected exposure to the agent to an exposure that is assumed not to be associated with toxic effects.

For oral or dermal exposures, the Average Daily Dose (ADD) was compare to a Reference Dose (RfD):

$$\text{H. Q.} = \text{Average Daily Dose / Reference Dose} \quad (7)$$

Any Hazard Quotient less than 1 is considered to be safe for a lifetime exposure.

2.8.2 Cancer risk

For chemicals that may cause cancer if ingested, risk is calculated as a function of Oral Slope Factor and can was calculated by using the formula:Risk = Oral Slope Factor * Lifetime Average Daily Dose (8)

2.8.3 Cross-media transfer equations used to generate exposure estimates

The formulae used to generate the contaminant exposure concentration in water were those described by the US-EPA (1990) for water to fish; vegetables; dairy and meat concentrations. The formula for the consumption of recreationally caught fish and shellfish-water to edible tissue is presented in equations below:

$$C(f) = BCF * (\frac{fat}{3}) * C(w) \quad (9)$$

$$\text{BCF} = [0.79 * \log(\text{Kow})] - 0.40 \quad (10)$$

Where:C (*f*) = concentration in fishC (w) = concentration in waterC(sd) = Concentration in Sediment

DN = Sediment Density (Relative to Water Density of 1.0 kgl^{-1}) (1.90)
OC = Organic Carbon Fraction of Sediment (4.00 %)
Koc = Octanol-Carbon Partition Coefficient of the Compound
Kow = Octanol - Water coefficient of the compound
BCF = Bioconcentration factor

2.8.4 Homegrown fruit and/or vegetables – Water to root (root uptake)

Limitations: not applicable to polar species, where RCF(w) = 0.82

$$C(r) = RCF\,(w) * C(w) \tag{11}$$

$$\log(RCF(w) - 0.82) = (0.77 * \log(Kow)) - 1.52 \tag{12}$$

Where:

C(r) = Concentration in root Calculated
C(w) = Concentration in water Chemical Specific
Proportion in root 100.00 %
RCF = Root concentration factor
Kow = Octanol-water partition coefficient of the compound Chemical Specific

2.8.5 Homegrown fruit and/or vegetables – Water to transpiration stream (root uptake)

$$C(st) = TSCF(w) * C(w) \tag{13}$$

$$TSCF(w)= 0.784 \exp -((\text{Log}(Kow) - 1.78) *2)/2.44) \tag{14}$$

Where:

C(st) = Concentration in Stem Calculated
C(w) = Concentration in Water Chemical Specific
Proportion in Stem: 100.00 %
TSCF = Transpiration Stream Concentration Factor
Kow = Octanol-water partition coefficient of the compound Chemical Specific

2.8.6 Homegrown meat or dairy – Water to edible tissue

Limitations: If either of these conditions occur, BCF = 0, Log Kow < 3.5, Log S > 4

where S is the water solubility of the compound.

$$C(t) = BCF(f) * F * C(w) \tag{15}$$

$$\log(BCF(f)) = -3.457 + 0.5\,(\log(Kow)) \tag{16}$$

Where:

C(t) = Concentration in edible tissue calculated

C(w) = Concentration in water chemical Specific
F = Fat content in tissue (dairy) 4.00 %
F = Fat content in tissue (Meat) 14.00 %
BCF = Bioconcentration factor for tissue fat chemical specific
Kow = Octanol-water partition coefficient of the compound chemical Specific

2.8.7 Exposure parameters used to calculate exposure estimates

The dose estimates in this assessment, as well as the risk estimates derived from them, refers only to the specific exposures that have been described in Table 3. The average daily dose was calculated taking into account the concentration of the chemicals in water, sediment, for a 70 Kg adult, assuming an intake of 0.054 kg fish on a daily basis (equivalent to 378 g per week). A range of risks is presented making use of average and 95th percentile concentrations of chemicals detected in water, calculated to represent concentrations expected in fish. The 95th percentile represents the "reasonable maximum" risk.

Exposure parameter	Amount
Events per year	350
Kg fish per day	0.054
Kg dairy	0.4
Kg meat per day	0.1
L water per day	2
Body weight	70 kg
Exposure duration	30 years

Table 3. Exposure parameters used to generate exposure estimates.

3. Result and discussion

The LOD of each compound for the analytes was determined as three times the standard deviation of seven independent replicate analyses. LOQs were determined as 3.3 times of LODs. Instrument detection limits ranged from 0.6 μgl^{-1} (DEHP) to 3.16 μgl^{-1} (4-NP) and the LOQs varied from 1.9 μgl^{-1} (DEHP) to 10.44 μgl^{-1} (4-NP) as presented in Table 4. The LODs and LOQs values are adequate for environmental monitoring of the target compounds and low enough compared to previous work on the analytes of interest (Fatoki and Noma, 2002; Yuan *et al.*, 2002; Cortazar *et al.*, 2005; Zhou *et al.*, 2005; Kayali *et al.*, 2006; Ling *et al.*, 2007) taking into account the complexity of the samples and the low sample amounts used. For wastewater and river samples, the LODs achieved in the present work were at similar levels or lower than those obtained in previous studies with GC–MS (Yuan *et al.*, 2002; Cortazar *et al.*, 2005; Kayali *et al.*, 2006). The chromatogram of the derivatized phenols and phthalate esters congeners are presented in Figure 3.

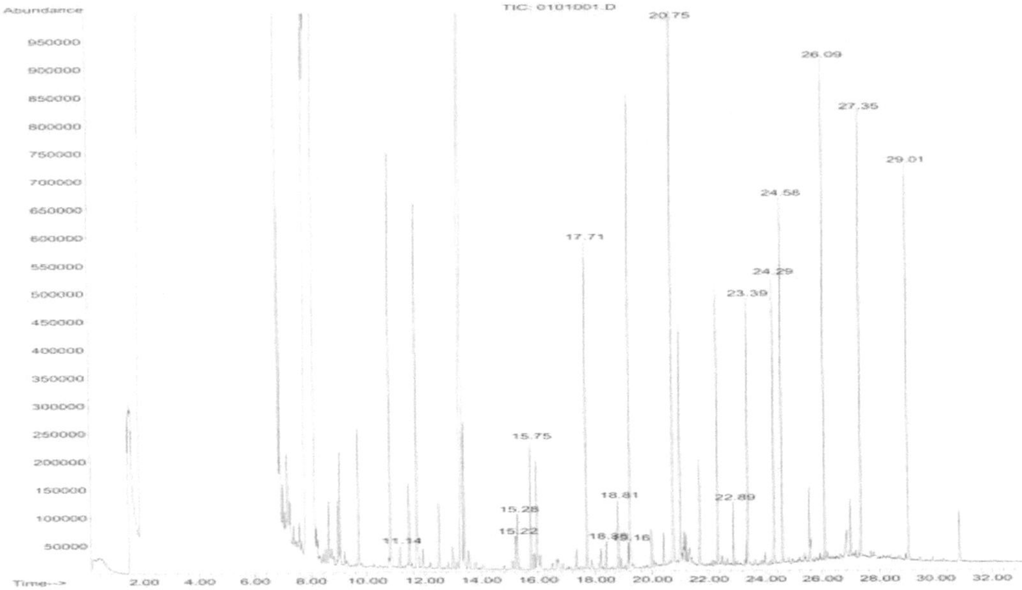

Fig. 3. Chromatogram of derivatized phenols and phthalate esters.

Five point calibration curves were constructed using triplicate injections of the derivatized standard. The retention time, target ion monitored, and the SPE recovery of the selected phenols and phthalates are presented in Table 4. Analysis of the result demonstrated the concordance of the response with a linear model as shown in Table 4, where the regression coefficient ranges from 0.976 to 1.000. The method precision and accuracy were satisfactory. The detectable concentration range was from 2.5 to 1000 μgl^{-1}. Due to non-availability of reference materials, the validation of the analytical method for extraction and elution was assessed through the recovery of standard mixtures of the target analytes in Milli-Q water. For the efficient quantification of the target compounds, analysis was performed within the linear portion of the calibration curve.

3.1 Health risk assessment

There are many associated adverse health effects if people are exposed to these chemical contaminants in excess doses. Where possible the study looked at whether people might be exposed to excessive concentrations through various pathways, such as if water were used for domestic purposes, if the water were used to irrigate vegetables, if fish living in the water were eaten on a regular basis, if the rivers were used for recreational swimming and lastly if meat were consumed from the area making use of the water. The classic example of a population that differs from the norm is subsistence fishers, who may consume as much as 10 times the amount of freshwater fish that most citizens do.

This population is of particular concern when evaluating surface water contamination in areas that are economically depressed or if the immune systems of the people in the area are compromised. The methodology used to asses this potential human health risk was that described by the US-EPA (1988, 1996) and the WHO (2002), making use of the risk assessment programme, Risk Assistant ™ (Thistle Publishers, 1996). DEHP and DBP were the only organic chemicals of those tested that could be included in the quantitative health risk assessment.

Compound	Retention Time (min)	Target ion (m/z)	Reference ion (m/z)	SPE Recovery (%)	LOD (μgl^{-1})	LOQ (μgl^{-1})	Correlation Coeffient R^2
Phenol	11.14	151	208	93.43 ± 0.05	2.2	7.18	1.000
2-CP	15.21	185	149, 93	98.21 ± 4.38	1.9	6.34	0.988
DMP*	15.27	163	77	83.72 ± 6.03	2.2	7.43	0.993
2,4-DMP	15.74	179	163, 149, 105	98.69 ± 8.43	1.4	4.78	0.987
4-C,3MP	17.71	199	93	76.21 ± 5.28	2.96	9.77	0.989
DEP*	18.38	149	177, 104, 77	98.46 ± 11.31	1.58	5.22	0.993
2,4-DCP	18.81	219	183, 125,93	94.1 ± 7.16	1.11	3.66	1.000
2-NP	19.15	196	180, 151, 136, 91	95.39 ± 11.68	1.36	4.47	1.000
4-NP	20.74	196	150, 135	88.19 ± 10.29	3.16	10.44	0.999
2,4,6-TCP	20.76	255	217, 159, 93	73.21 ± 0.05	2.81	9.63	0.999
DBP*	22.89	149	207	98.99 ± 8.27	0.9	2.9	0.978
2,4-DNP	23.39	241	225, 195, 137	96.34 ± 2.93	1.63	5.36	0.986
2-M, 4,6-DNP	24.29	255	239, 209 179, 149	90.33 ± 6.18	1.48	4.87	0.976
PCP	24.58	323	93	92.64 ± 11.39	2.23	7.37	0.998
BBP*	26.09	149	206, 91	97.43 ± 18.31	0.6	2.9	0.987
DEHP*	27.35	149	279, 167	101.32 ± 0.21	0.6	1.9	0.989
DOP*	29.01	149	279, 57	90.77 ± 5.39	1.41	4.65	0.988

*Compound not affected by MTBSTFA derivatization

Table 4. Retention time, target ion, limits of detection and quantification in GC-MS of the selected phenols and phthalates recoveries (n = 7).

The average concentrations detected in all the sample sites over the sampling period of a year was used as a most likely scenario to determine what risks (if any) were involved as a screening risk assessment. If a chemical was found to be responsible for risks considered by the US-EPA and WHO to be unacceptably high, a more detailed assessment for that chemical was investigated, making use of the spread of the data, averages, and identifying which sampling site was responsible for the highest concentrations detected. The following graphs (Figures 4 & 5) illustrates the average concentrations of the chemicals detected at the sampling sites used in the primary screening for human health risk assessment.

DBP was found at highest concentrations in both river water samples and wastewater effluents, followed by nitro-phenol (NP) and DEP (Figures 4 & 5). Human dose-response data was available for DEHP and DBP to allow a quantitative health risk assessment to be performed (ATSDR, 1995; 2001; 2002). The results of the exposure calculations are given in the Table 5 and are presented as both Average Daily Dose (ADD) and Lifetime Average Daily Dose (LADD) in mg/kg/d.

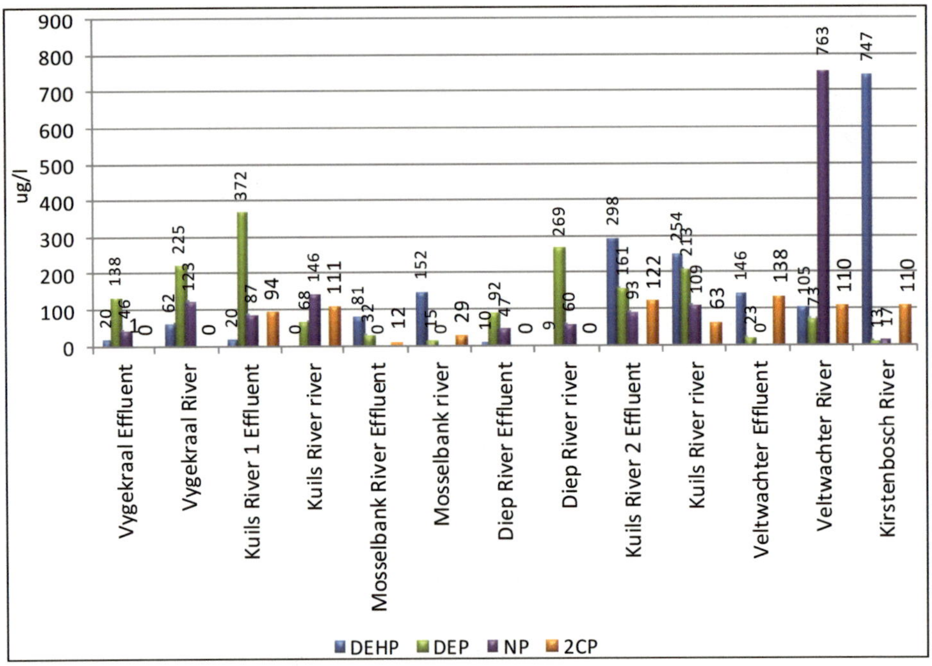

Fig. 4. Concentrations (μgl-1) of phthalate and phenolic congeners detected in river and WWTP effluents at different sites (excluding DBP).

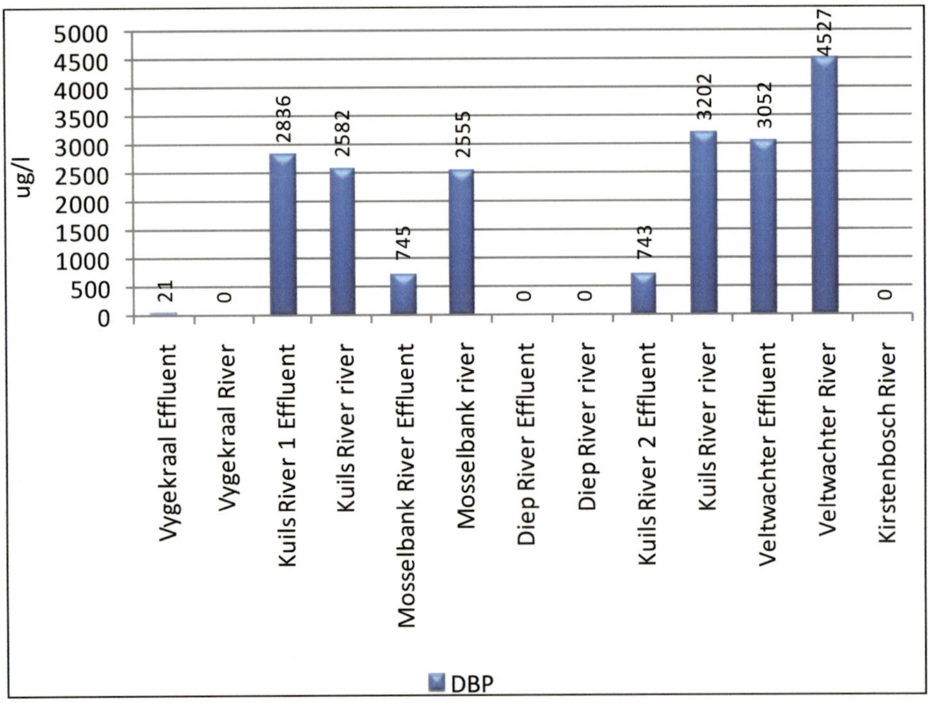

Fig. 5. DBP concentrations (μgl-1) detected in effluent and river water samples at the different sites.

Based on the exposure assumptions described in the section above, risks of developing cancer and toxic effects were calculated for the various phthalate chemicals where sufficient data was available. Most of the chemicals were found at concentrations to be below those where "unacceptable" risks, as defined by both the WHO and US-EPA, are anticipated. However, risks of developing cancer may be as high as 2 in one thousand resulting from exposure to DEHP (Figure 6) resulting predominantly from exposure through vegetables that have been irrigated with the contaminated water, and to a lesser extent, through the consumption of fish, grown in the contaminated water.

DEHP was detected at high concentrations at Kirstenbosch, Kuils River, Mosselbank River and Vygekraal River. In general, river waters contained higher concentrations than treated effluents of the waste water treatment works (Figure 6). This risk would result if the water were used to irrigate vegetables or if fish grown in the water were consumed on a regular basis.

Site	Chemical	ADD (mg/kg/d)	LADD (mg/kg/d)
Vygekraal River	DEPH	0.02474	0.0106
	DBP	0	0
Vygeraal Effluent	DEHP	0.007983	0.0003421
	DBP	0.003242	0.001389
Kuils River 1	DEHP	0	0
	DBP	0.3895	0.1708
Kuils River (1) Effluent	DEHP	0.007982	0.003421
	DBP	0.4377	0.1876
Mosselbank River	DEHP	0.06066	0.0206
	DBP	0.3943	0.169
Mosselbank Effluent	DEHP	0.03233	0.01385
	DBP	0.115	0.04927
Diep River	DEHP	0.003592	0.001539
	DBP	0	0
Diep River Effluent	DEHP	0.003991	0.00171
	DBP	0	0
Kuils River (2)	DEHP	0.1014	0.0435
	DBP	0.4941	0.2118
Kuils River (2) Effluent	DEHP	0.1189	0.05097
	DBP	0.1147	0.04914
Veldwachter River	DEHP	0.04191	0.1796
	DBP	0.6986	0.02497
Veldwachter Effluent	DEHP	0.05827	0.02497
	DBP	0.471	0.2019
Kirstenbosch Stream	DEHP	0.1278	0.1278
	DBP	0	0

Vygekraal Effluent = Athlone WWTP effluent; Kuils River (1) Effluent = Bellville WWTP Effluent; Mosselbank Effluent = Kraaifontein WWTP Effluent; Kuils River (2) Effluent = Zandvliet WWTP Effluent Veldwachter Effluent = Stellenbosch WWTP Effluent; Kirstenbosch Stream = Control Site.

Table 5. Predicted total average daily doses and lifetime average daily doses, based on average concentrations of phthalates.

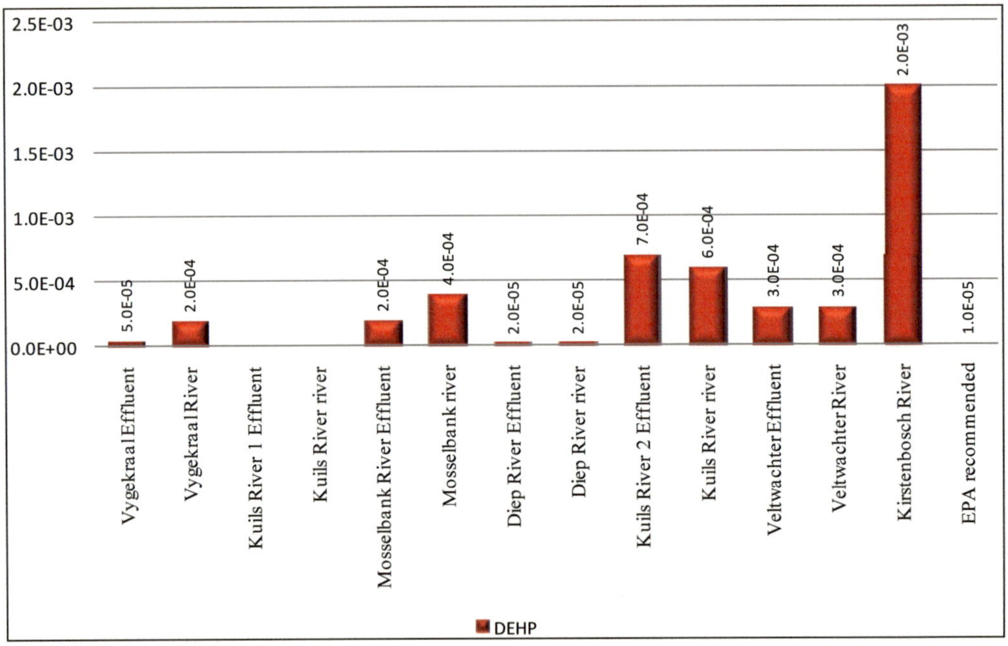

Fig. 6. Cancer risks from DEHP exposure.

Toxic risks could be anticipated resulting from exposure to both DEHP and DBP with individual exposure concentrations predicted at up to 14 times that considered to be safe for a lifetime exposure (Figure 7 & 8). However, the certainty of the reference dose, or the dose considered to be safe, has a safety factor of 100 built into it for both DEHP and DBP (ATSDR, 2002; and ATSDR, 2001 respectively). The safety factors built into the reference doses for DEHP and DBP are to allow for extrapolation from animals to humans (a factor of 10) and to allow for variability within humans (another factor of 10) (ATSDR 2001 & 2002). The predicted risks indicate that a possible risk exists and does not indicate a definite risk as the exposures are modelled and not based on actual measurements.

The driver of the human health risk was identified through this exercise. The chemicals responsible for the risks include DEHP and to a lesser extent, DBP (Figures 6 & 7). DEHP was found to be the major contributor of risk of developing cancer in this screening health risk assessment. The highest potential risks were observed at Kirstenbosch resulting from DEHP detected in the river water. The potential risk through the use of this water is if it were used to irrigate vegetables.

This section examined whether possible human health effects might be anticipated based on chemical contaminants detected in wastewater effluents and in rivers throughout the Western Cape, South Africa. In order to determine whether this is possible, a human health risk assessment was conducted by modelling the chemical contaminant concentrations expected in vegetables, fruit, fish and meat based on levels detected in water. Trans-media calculations (water to fish; water to fruit and vegetables and water to meat) were conducted based on individual chemical parameters described in the earlier sections.

The screening risk assessment identified the chemicals that could be responsible for adverse health effects if drinking the untreated water or eating fish , fruit, vegetables or meat , over a 30 year period were to occur. Although not present at the highest concentrations, the chemicals that were of principal concern were identified as DEHP and to a lesser degree, DBP and arsenic. The type of adverse effect that might result was also identified as predominantly carcinogenic, with possible reproductive system toxic effects being anticipated, as the predicted doses were well below those considered safe by the WHO and US EPA.

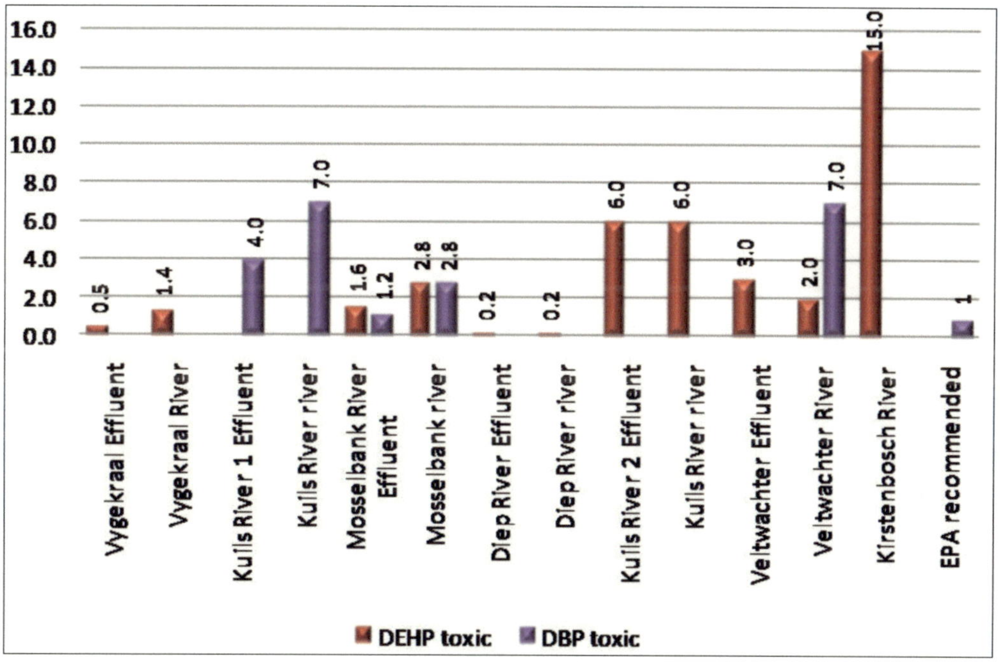

Fig. 7. Hazard quotients for individual phthalates.

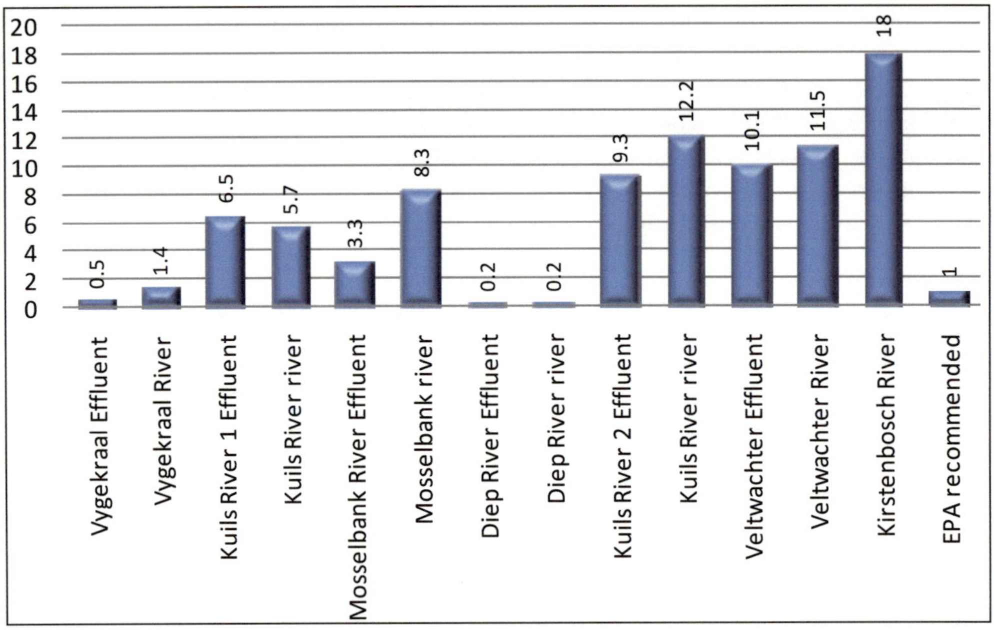

Fig. 8. Hazard quotients for total phthalates.

4. Conclusion

This screening risk assessment has highlighted that possible health risks can be anticipated resulting from ingestion of vegetables irrigated with the water and ingestion of fish from the rivers on a regular basis. There are many uncertainties in any health risk assessment, and this study presents a screening or rapid human health risk assessment. Seasonal and spatial variations were considered in this health risk assessment as the average concentrations tested over the 4 seasons were used in the average daily dose calculations. In addition to sample variation, dose calculations also represent uncertainty, based on the assumption of the number of times a year that people eat certain foods and the amount of that food eaten. Future investigations need to focus on verifying the uptake of phthalates into vegetables and fish via water as this has highlighted that although levels were considered to be safe in the water, bio-accumulation is possible into both fish and vegetables to levels considered to be unacceptable by the US EPA and WHO.

5. References

Ahel, M., Schaffner, C. and Giger, W. (1996). Behaviour of alkylphenol polyethoxylate surfactants in the aquatic environment —III. Occurrence and elimination of their persistent metabolites during infiltration of river water to groundwater, *Water Research* 30:37–46.

Ahel, M., Giger, W. and Koch, W. (1994). Behaviour of alkylphenol polyethoxylate surfactants in the aquatic environment-I. Occurrence and transformation in sewage treatment. *Water Research* 28: 1131-1142.

ATSDR, (1995). Agency for Toxic Disease Registry. Toxicological profile for Diethyl phthalate. Public Health Service Agency for Toxic Substances and Disease Registry. http://www.atsdr.cdc.gov/toxprofiles/TP.asp?id=603&tid=1120, accessed 23 August 2010)

ATSDR (2001) Toxicological profile Di-n-butyl phthalate. Public Health Service Agency for Toxic Substances and Disease Registry. (http://www.atsdr.cdc.gov/toxprofiles/TP.asp?id=603&tid=1120, accessed 23 August 2010)

ATSDR (2002). Toxicological profile for D2-(2 ethylhexyl) phthalate. . Public Health Service Agency for Toxic Substances and Disease Registry. (http://www.atsdr.cdc.gov/toxprofiles/TP.asp?id=603&tid=1120, accessed 23 August 2010)

Awal, M.A., Kurohmaru, M. Ishii, M., Andriana, B.B., Kanai, Y. and Hayashi, Y. (2004). Mono(2-ethyl hexyl) phthalate (MEHP) induces spermatogenic cell apoptosis in guinea pig testes at prepubertal stage in vitro. *International Journal of Toxicology* 23: 349–355 Backeberg, G.R. (1996). Presidential Address: Constitutional change and natural resource policy reform in South Africa. *Agrekon,* 35 (4), 160-169.

Bennion, H., Hilton, J., Hughes, M., Clark, J., Hornby, D., Fozzard, I., Phillips, G. and Reynolds C. (2005).The use of a GIS based inventory to provide a national assessment of standing waters at risk from eutrophication in Great Britain. *Science of the Total Environment* 344: 259-273.

Blignaut, J. and J. Heerden, (2009). The impacts of water scarcity on economic development initiatives. Water SA 35 (4): 415-420

Cortazar, E., Bartolome, L., Delgado, A., Etxebarria, N., Fern´andez, L.A., Usobiaga, A. and Zuloaga, O. (2005). Optimisation of microwave-assisted extraction for the determination of nonylphenols and phthalate esters in sediment samples and comparison with pressurised solvent extraction. *Analytica Chimica Acta* 534: 247-254.

De Sousa, J. P. B., da Silva Filho, A. A., Bueno, P. C. P., Gregório, L. E., Furtado, N. A. J. C., Jorgea, R. F. and Bastosa, J. K. 2009. A validated reverse-phase HPLC analytical method for the Quantification of Phenolic compounds in *Baccharis dracunculifolia. Phytochemical Analysis* 20: 24-32.

Fatoki, O.S., Bornman, M., Ravandhalala, L., Chimuka, L., Genthe, B. and Adeniyi, A. (2010). Phthalate ester plasticizers in freshwater systems of Venda, South Africa and potential health effects. *Water SA* 36(1): 117-126.

Fatoki, O.S. Awofolu, O.R. and Genthe, B. (2004). Cadmium in the Umtata River and the associated health impact on rural communities whose are primary users of water from the river. *Water SA* 30(4):507-513.

Fatoki, O.S. and Noma, A. (2002). Solid phase extraction method for selective determination of phthalate esters in the aquatic environment. *Water, Air and Soil Pollution* 140: 85-98.

Fatoki, O.S., Muyiwa, N.Y.O. and Lujiza. (2001). Situation analysis of water quality in the Umtata River catchment. *Water SA* 27 (4): 467 - 474.

Huang, J., Wang, H., Jin, Q., Liu, Y. and Wang, Y. (2007). Removal of phenol from aqueous solution by adsorption onto OTMAC-modified attapulgite. *Journal Environmental Management* 84: 229-236.

Jackson, V. A., Paulse, A. N., van Stormbroek, T., Odendaal, J. P. and Khan, W. (2007). Investigation into metal contamination of the Berg River, Western Cape, South Africa. *Water SA* 33(2): 175-182.

Jackson, V.A., Paulse, A.N., Odendaal, J.P. and Khan, W. (2009). Investigation into the metal contamination of the Plankenburg and Diep Rivers, Western Cape, South Africa. *Water SA* 35 (3): 289-299.

Kayali, N., Tamayo, F.G. and Polo-Diez, L.M. (2006). Determination of dimethylhexyl phthalate in water by solid phase microextraction coupled to high performance liquid chromatography. *Talanta* 69: 1095-1099.

Komnenic, V., Ahlers, R. & van der Zaag, P. (2009). Assessing the usefulness of the water poverty index by applying it to a special case: Can one be water poor with high levels of acess. *Physics and Chemistry of the Earth* 34: 219–224.

Lambrot R, Muczynski V, Lecureuil C, Angenard G, Coffigny H, Pairault C. (2009). Phthalates impair germ cell development in the human fetal testis in vitro without change in testosterone production. *Environmental Health Perspective* 117:32–37.

Latini, G., Wittassek, M., DelVechio, A., Presta, G., De Felice, C. and Angerer, J. (2009). Lactational exposure to phthalates in southern Italy. *Environment International* 35: 236-239.

Ling, W., Gui-bin, J., Ya-qi, C., Bin, H., Ya-wei, W. and Da-zhong, S. (2007). Cloud point extraction coupled with HPLC-UV for the determination of phthalate esters in environmental water samples. *Journal of Environmental Sciences* 19: 874–878.

Llompart, M., Lourido, M., Landın, P., Garcıa-Jares, C. and Cela, R. 2002. O ptimization of a derivatization solid-phase microextraction method for the analysis of thirty phenolic pollutants in water samples. *Journal of Chromatography A* 963: 137-148.

Madungwe, E. & Sakuringwa, S. (2007). Greywater reuse: A strategy for water demand management in Harare. *Physics and Chemistry of the Earth*, 32, 1231-1236.

Malley, Z.J.U., Taeb, M., Matsumoto, T. & Takeya, H. (2009). Environmental sustainability and water availability: Analyses of the scarcity and improvement opportunities in the Usangu plain, Tanzania. Physics and Chemistry of the Earth 34: 3–13.

Marcucci, M. & Tognotti, L. (2002). Reuse of wastewater for industrial needs: the Pontedera case. *Resources, Conservation and Recycling* 34: 249-259.

Mylchreest E, Madhabananda S, Sar M, Cattley RC, Foster PMD. (1999). Disruption of androgen regulated male Reproductive development by Di(n-butyl)phthalate during late gestation in rats is different from flutamide. Toxicol Appl Pharmacol. 156:81-95.

NRC. 1983. Risk Assessment in the Federal Government: Managing the Process. National Research Council, National Academy Press. Washington, DC, USA. NRC. 1993. Issues in Risk Assessment. National Research Council, National Academy Press. Washington, DC, USA.

Olujimi, O. O., Fatoki, O.S, and Odendaal, J.P. (2011b). Method development for simultaneous determination of phthalate and eleven priority phenols as tert-butyldimethylsilyl derivatives in grab samples from wastewater treatment plants using GCMS in Cape Town, South Africa. *Fresenius Environmental Bulletin* 20 (1): 69-77

Olujimi, O. O., Odendaal, J.P., Okonkwo, O.J. and Fatoki, O.S. (2011a). Solid-phase extraction method for the analysis of eleven phenolic priority pollutants in water samples using a newly launched column. *Asian Journal of chemistry* 23 (2): 657-662

Oweis, T. & Hachum, A. (2009). Optimizing supplemental irrigation: Tradeoffs between profitability and sustainability. *Agricultural Water Management* 96: 511–516.

Qiao, G., Zhao, L. & Klein, K.K. (2009). Water user associations in Inner Mongolia: Factors that influence farmers to join. Agricultural Water Management 96: 822–830.

Risk Assistant™ (1995. The Hampshire Research Institute Inc. Thistle Publishing, Alexandria VA, USA.

Spinks, A.T., Dunstan, R.H., Harrison, T., Coombes, P. & Kuczera, G. (2006). Thermal inactivation of waterborne pathogenic and indicator bacteria at subboiling temperatures. *Water Research* 40: 1326-1332.

Sexton, K., Callahan, M.A., Bryan, E.F. 1995. Estimating exposure and dose to characterize health risks: The role of human tissue monitoring in exposure assessment. *Environmental Health Perspectives* 103: 13-29.

Sun, Q., Deng, S., Huang, J., Shen, G. and Yu, G. (2008). Contributors to estrogenic activity in wastewater from a large waterwater treatment plant in Beijing, China. *Environmental Toxicology and Pharmacology* 25: 20-26.

Vethaak, A.D., Lahr, J., Schrap, S.M., Belfroid, A.C., Rijs, G.B.J., Gerritsen, A., Boer, J.D. Bulder, A.S., Grinwis, G.C.M., Kuiper, R.V., Legler, J., Murk, T.A.J., Peijnenburg, W., Verhaar, H.J.M. and Voog, P.D. (2005). An integrated assessment of estrogenic contamination and biological effects in the aquatic environment of the Netherlands. *Chemosphere* 59: 511–524.

Vo, T.T.B., Jung, E.M., Dang, V.H., Yoo, Y.M., Choi, K.C. and Yu, F.H.. 2009. Di-(2 ethylhexyl) phthalate and flutamide alter gene expression in the testis of immature male rats. Reprod Biol Endocrinol 7: 104–118.

Voutsa, D., Hartman, P., Schaffner, C. and Giger, W. (2006). Benzotriazoles, Alkylphenols and Bisphenol A in Municipal Wastewaters and in the Glatt River, Switzerland. *Environmental Science Pollution Research* 13 (5): 333–341.

Wang, Y., Hu, W., Cao, Z., Fu, X. and Zhu, T. (2005). Occurrence of endocrine-disrupting compounds in reclaimed water fromTiajin, China. Anal. Bioanal Chem 383:857-863.

Ying, G.G., Kookana, R.S. and Ru, Y.J. (2002). Occurrence and fate of hormone steroids in the environment. *Environment International* 28: 545–551.

Yuan, B., Li, Z., and Graham, N. (2008). Aqueous oxidation of dimethyl phthalate in a Fe(VI)-TiO_2-UV reaction system. *Water Research* 42: 1413-1420.

Yuan, S.Y., Liu, C., Liao, C.S. and Chang, B.V. (2002). Occurrence and microbial degradation of phthalate esters in Taiwan river sediments. *Chemosphere* 49: 1295-1299.

Zhou, F., Li, X. and Zeng, Z. (2005). Determination of phenolic compounds in wastewater samples using a novel fiber by solid-phase microextraction coupled to gas chromatography. *Analytica Chimica Acta* 538, 63-70.

Zuccato, E., Castiglioni, S., Fanelli, R., Reitano, G., Bagnati, R., Chiabrando, C., Pomati, F., Rossetti, C. and Calamari, D. (2006). Pharmaceuticals in the Environment in Italy: Causes, Occurrence, Effects and Control. *Environmental Science Pollution Research* 13 (1), 15–21.